Knowledge Economies

Today, concerns about national competitiveness and economic development are closely linked to notions of the information society and knowledge-based economies. Some see the emergence of a 'new economy', which is based on information, communication, media and biotechnologies. As in previous bursts of economic growth, these innovative industries emerge and grow in specific geographic locations, now called 'clusters'. The well-known case of Silicon Valley was merely the first of a large number of such clusters to have developed in recent years. It is argued that clusters are characterized by cooperative and competitive, trustful and rivalrous, exchange and favour-based business interactions.

This book traces the theoretical explanation for clusters back to the work of classical economists and their more modern disciples who saw economic development as a process involving serious imbalances in the exploitation of resources. First, natural resource endowments explained the formation of nineteenth- and early twentieth-century industrial districts. Today geographical concentrations of scientific and creative knowledge are the key resource. But these require a support system, ranging from major injections of basic research funding, to varieties of financial investment and management, and specialist incubators for economic value to be realized. These are also specialized forms of knowledge that contribute to a serious imbalance in the distribution of economic opportunity.

The key question is whether the techniques of cluster building can be learned and promotional policies implemented to offset the natural imbalances in the distribution of these specialized knowledge resources. Developing on the idea of multi-level governance and policy-making, *Knowledge Economies* reviews cases where national governments working intelligently with regional, local and, in Europe, supra-national governance organizations, have been able to implant clusters in places that previously did not have them. Learning about the nature of clusters, and from experiences in developing them with the help of policy intervention, will assist the process of strengthening existing ones, developing new ones and revitalizing older ones. In the process, the goals of regional equity and competitiveness should be enhanced.

Philip Cooke is Professor of Regional Development and Director of the Centre for Advanced Studies at the University of Wales Cardiff. He is also editor of the inter-disciplinary journal, *European Planning Studies.*

Routledge Studies in International Business and the World Economy

Knowledge Economies

Clusters, learning and
cooperative advantage

Philip Cooke

London and New York

First published 2002
by Routledge
2 Park Square, Milton Park, Abingdon, Oxfordshire, OX14 4RN

Simultaneously published in the USA and Canada
by Routledge
711 Third Avenue, New York, NY 10017

First issued in paperback 2014

Routledge is an imprint of the Taylor and Francis Group, an informa company

Transferred to Digital Printing 2006

Typeset in Galliard by Taylor & Francis Books Ltd

British Library Cataloguing in Publication Data
A catalogue record for this book is available from the British Library

Library of Congress Cataloging in Publication Data
Cooke, Philip.
Knowledge economies: clusters, learning and cooperative advantage/
Philip Cooke.
Includes bibliographical references and index.
1. Knowledge management. 2. Business incubators. 3. Natural resources.
4. Evolutionary economics. 5. Economic development. I. Title.

HD30.2 .C657   2001
330.1–dc21   2001034877

ISBN 13: 978-0-415-16409-2 (hbk)
ISBN 13: 978-0-415-75716-4 (pbk)

Contents

Illustrations

Tables

Figures

Acknowledgements

The idea for this book grew from a commission from the National Economic and Social Council in Dublin who asked me to conduct a study of economic performance in dynamic growth regions in Europe. The findings of that study were published in NESC's report on *Networking for Competitive Advantage*. As no book-length treatment of the themes dealt with existed at the time, I proposed what has, over the intervening years turned into a somewhat different outcome, to Routledge who agreed to publish a thematically structured book on learning, networking and clustering in regional economies. My first thanks go to Rory O'Donnell, at the time director of NESC, and his board, for asking me to do the initial research and helping it, along with Noel Cahill, to fruition. I also found David Jacobson a willing ally in this and many subsequent research tasks. At Routledge, Petra Recter and, more recently, Ann Michael have been patient publishing editors.

Because of some other research and publishing commitments, this project moved rather more slowly than expected, but I benefited from advice and comments from, among others, Roland Sturm, Bjørn Asheim, Meric Gertler, David Wolfe, Anders Malmberg and Alaric Maude. The project started to get going in earnest following an invitation, marvellously administered by Alaric, from the Autralasian Regional Science Association to talk at their annual conference in Barossa in 1998. A hastily organized lecture tour through Sydney, Melbourne and Adelaide caused me to frame Chapters 4 to 6 as Power Point presentations, although much of the detailed reportage did not exist at that time, only structures. I am grateful to Bob Fagan, John Langridge, Kevin O'Connor, Margo Huxley and Andrew Beer, amongst many others in Australia, for showing interest and giving good advice on the project. In the UK, I also valued discussions about social capital with Rich Meegan and clusters with Martin Boddy and Jim Simmie during the ESRC 'Cities' research programme meetings. In Northern Ireland, Aidan Gough and Angela McGowan from the Northern Ireland Economic Council were valued collaborators.

Helping me get to and through the last two chapters were a number of other friends. In Germany, Ludwig Schätzl and Javier Revilla Diez persuaded me to do a summer school at the Economic Geography department in Hannover University. More Power Points and steps into the only dimly perceived reaches of the New Economy. John Rees, who used to live in Richardson, Texas one of the key cases in the book, helped put me right on some historical points, as did Vanessa Weigelt from Hannover who produced some outstanding maps for the Richardson Chamber of Commerce and her master's thesis, one of which she has been kindly allowed to be reproduced here. My old friends, Goio Etxebarria, Mikel Gomez Uranga and their colleagues helped me out enormously with some of the institutionalist economic theory in the book. Another old friend, Glyn Williams, has been stimulating to talk to about the Digital Economy, on which he is now an expert adviser to the European Commission while a new friend Frédéric Richard of UNIDO made me think and learn how all this impacted upon less developed countries. It was also a major experience to work on UK Minister of Science Lord Sainsbury's Biotechnology Clusters Task Force, and some of the background for that report from Massachusetts and the UK has been drawn upon, though all interpretations are those of this author alone. Finally, my former student, Nick Henry, who has so distinguished himself by his detailed work on the UK motor sport cluster, must be thanked for reading the manuscript and making excellent suggestions for improvement, as also did my old friend Rich Florida. Thanks to everyone for helping me, through their knowledge and learning, realize the virtues of cooperative advantage.

Philip Cooke
Cardiff
March 2001

The author and publishers would like to thank the following for granting permission to reproduce figures and tables in this work:

The Centre for Advanced Studies, Cardiff University, for figure 6.1 and table 3.1, from a report on 'Cambridgeshire Training and Enterprise Council'. (Figure 6.2 and table 6.6 in this book.) The Department of Economics, University of Keele, for chart 1 (2) 'Flow chart for the manufacture of earthenware showing dependence upon supporting industries' from *The Production Organisation of the Ceramics Industry*, by Machin & Smyth, 1969. (Figure 7.2 in this book.) Dirke Dohse for the table 'Firm Assessments of BioRegio contest', from *The BioRegio Contest*, Institute of World Economics (mimeo), 1999. (Table 7.2 in this book.) The Northern Ireland Economic Council for the figure of 'Northern Ireland's Partial Innovation System Pyramid', from page 65 of the RIS report, 2001. (Figure 3.2 in this book.) Richardson Chamber of Commerce and Vanessa Weigelt for the figure of 'Telecommunication Corridor, Richardson, Texas', from www.richardson.com, 2000. (Figure 6.1 in this book.)

Every effort has been made to contact copyright holders for their permission to reprint material in this book. The publishers would be grateful to hear from any copyright holder who is not here acknowledged and will undertake to rectify any errors or omissions in future editions of this book.

Abbreviations

BMBF	Ministry of Education, Science, Research and Technology (Germany)
BTH	Biotechnology Centre Heidelberg
CAE	Cambridge Advanced Electronic
CNOC	Company Network Operations Command Centre
DKFZ	(Helmholtz) Cancer Research Centre
DTI	Danish Technological Institute
ERBI	Eastern Region Biotechnology Initiative
ERP	Enterprise Resource Planning
EU	European Union
FE	further education
FISPA	Finnish Science Park Association
HE	higher education
HI	Heidelberg Innovation GmBH
ICT	Information and Communication Technology
IDB	Industrial Development Board
IPO	initial public offering
IPR	intellectual property rights
IRTU	Industrial Research and Technology Unit
IZB	Biotechnology Innovation Centre
KP	Kleiner Perkins
LEDU	Local Economic Development Unit
MDA	Multimedia Development Association
MIT	Massachusetts Institute of Technology
MMI	Molecular Machines and Industries
MNC	multinational corporation
MPI	Max Planck Institute
NEIS	New Economy Innovation System
NIC	Newly Industrialized Country
NISPF	Northern Ireland Science Park Foundation
NIST	National Institute of Standards and Technology
NRW	North Rhine-Westphalia
OECD	Organization for Economic Cooperation and Development

PCB	printed circuit board
PCI	Italian Communist Party
POP	point of presence
PSI	Italian Socialist Party
R&D	research and development
RIS	Regional Innovation System
RITTS	Regional Innovation and Technology Transfer Strategies
RLB	Rover Learning Business
ROI	Republic of Ireland
ROW	Rest of World
RTZ	Rechtrheinisches Technologie Zentrum
SBIR	Small Business Innovation Research
SME	small and medium-sized enterprise
SONET	Synchronous Optical Network
TBC	Technology Business Council
TEC	Training and Enterprise Council
TEKES	Finnish Technology Development Centre
TI	Texas Instruments
TMT	technology, media and telecoms
TPZ	Future Technology Programme
TTN	Technology Training Network
TVEP	Thames Valley Economic Partnership
VC	venture capital
VTT	State Technical Research Centre (Finland)

1 Clusters, collective learning and disruptive economic change

Introduction

This book explores a particular economic phenomenon of our time, the emergence and development of industry *clusters*, asking the question why this kind of industrial organization reappeared in the advanced economies after it had more or less disappeared in the mid-twentieth century. It sets that question in the framework of a much larger one that has troubled many development economists, regional scientists and politicians with a concern about social and geographical inequality since Malthus first asked Ricardo why some nations are rich and others poor. This larger question was investigated recently and not entirely satisfactorily answered by David Landes (1998) who explained it in terms of the presence or absence of a cultural will to 'live for work'. Tautology apart, to live for work entails some things that this book is also interested in, namely, capabilities of learning and innovation as key economic instruments. These two concepts are at the heart of the idea of a 'knowledge economy'. But what Landes and others, like Fukuyama (1995; 1999) regard as normal, that is, individualistic competition in an ordered, economic equilibrium where thrift and honesty are justly rewarded, the argument that develops over the next eight chapters takes as abnormal. The knowledge economy consists of fragmentary 'knowledge economies'. This is for three key reasons concerning, first, *disequilibrium* or economic and social imbalance, which is not presumed to be unusual but quite the contrary. Second, *collaborative* economic action, is presented here as the most important organizational aspect of modern capitalism, but also one that has been vital to market economies from the start, despite the presumption in much economics that only individuals matter. While, third, the *systemic* nature of strategic competitiveness in the capabilities of specific groups of private and public actors to produce and implement actions based on consensus is of more importance than individual opportunism.

These are not particularly original points of divergence from orthodoxy. They are shared widely among a wide range of more heterodox thinkers who are interested in the social economy and take an evolutionary perspective on economic change, influenced particularly by the ideas of Joseph Schumpeter about the causes of such change. As is well known, Schumpeter was interested in *entrepreneurship*, but became even more interested in understanding innovation as a process that certain kinds of entrepreneur facilitated. Hence, he also fell into the trap of focusing on disruptive economic change as an effect of heroic individual genius. This was the legacy of the influence of Nietzsche's philosophy on his work, as is shown in Chapter 2. That this never really left him is testified to by the displacement of entrepreneurial heroism from the individual innovator to the R&D engineer in the large corporation, where his studies in the USA led him to conclude the modern wellsprings of innovation lay. In this book the real sources of contemporary innovation are shown residing in neither the individual entrepreneur nor the research laboratories of large firms but in networks of social relationships between such organizations and others of consequence to the discovery being sought and commercialized. Revealing the circuitry of knowledge economies is a complex task because it means finding out how the processes of knowledge generation and transfer to the point of exploitation function.

Recent research on what are popularly called 'new economy' industries like Internet content provision show the importance of knowledge networks and the very high value within them of enterprise support contacts, notably varieties of investment manager, 'venture catalyst', and 'incubation' or early-stage venture capitalist (Cooke, 2000; Zook, 2000; Keeble and Nachum, 2001; Sternberg, 2001). These are valued most for their scarce management expertise, despite a common assumption that it is their investments that count most. High, localized correlations between such businesses and services outweigh those between dot.coms and scientific or technological labour. However, occasionally the circuitry can be illuminated by exploring how it doesn't function or ceases to function when it once did. Much of the research that helps to do this is discussed in Chapters 7 and 8. Proving the negative is more difficult than demonstrating the positive, and researching failed cases is far less glamorous or marketable than disclosing 'new industrial spaces' which is why there is less material to call on to explain failures. Nevertheless, the book explores some in the homeland of industry clusters, the 'industrial age' districts near Manchester and Birmingham in the UK.

What the book tries to show as convincingly as possible is that clusters are crucial to economic imbalance, that they rest upon collaboration of a generally non-market-destroying type that is simply essential for modern economic organization, and that clusters have systemic organizational characteristics that go against much economic orthodoxy. For example, in Norton's (2000) book on the 'new economy,' he draws on Micklethwait and Woolridge's (2000) book which summarizes the economy culture of Silicon Valley as conveying a sense of loyalty to the place rather than the firm. This is expressed in such practices as reinvestment in the community, collaboration and 'tolerance of treachery'. The last of these lends a certain Hobbesian flavour to the composition and provokes a query about what is often said to be a key character of clusters in such places, their high ratio of trust in business transactions. Yet it is consistent with the thesis that *knowledge is in the networks* because each move in the interactive innovation process requires learning from others than those involved in the preceding move. So dropping a partner, competing against them for a contract, but maybe returning to them for its implementation, or for a future contract bid are not seen as bad form. On the contrary, these are the means by which the wellsprings of creativity flow and a key source of the 'spillovers' (Anselin *et al.*, 1997) that knowledge economies need and clusters supply. Recognition of the need to reproduce that characteristic is captured in the practice of reinvesting individual wealth generated back into the community, often as business angel investment. But collaboration is a key means by which that wealth is accumulated in the first place.

This brings us to the geographical dimension of clustering for learning, knowledge transfer, collaboration and the exploitation of spillovers. The argument here is that clusters are geographically localized and this causes imbalances between local areas that have them and those that do not. This has repercussions upwards to regions within countries and between countries themselves when the clusters in question have sufficient economic weight. London's financial cluster and Silicon Valley's Information and Communication Technology (ICT) cluster have a disproportionate impact on the trade balances of the UK and the USA. Italy's cluster areas in its north-central belt are far more prosperous than the Mezzogiorno region where they do not exist, something which is reversed in Germany where the south with automotive and ICT clusters is richer than the north whose 'industrial age' clusters are in decline.

Because clusters are focused geographical settings where industry specialization occurs, they develop external economies of scope and scale that it was once thought only single, large firms could manage. Modern ICT assists the routine part of this such as transmission of software, databases, designs and other forms of codified knowledge. But proximity in a cluster offers the opportunity for tacit knowledge exchange or 'treacherous' learning that may be hindered in large firms by 'group-think' and corporate culture. This is what accounts for the observation by de Geus (1997) that the average age of most large firm identities is around forty years. Large firms that do not conform to that fate change themselves, like Nokia, and survive much longer. So, in general, under contemporary knowledge-intensive market and competitiveness conditions it pays to be in a cluster or to simulate the kind of synergies from corporate re-design and re-invention that cluster networks supply.

This brings us to a key point about knowledge economies and their definition. Clearly, all human economic activity depends upon knowledge so, in a trivial sense, all economies are 'knowledge economies'. But because knowledge cannot be possessed in the way, for example, gold can, it can be appropriated by anyone capable of using it. This is despite the fact that it must be protected by patents. These after all are mainly a means for securing some economic return to invention rather than keeping knowledge confidential. There are three key issues: first, knowledge ages and is superseded by new knowledge that ideally requires what Johnson (1992) calls 'creative forgetting', namely, the stowing away of redundant knowledge and the learning of new. This can be a long and painful process, an illuminating, failed example of which is given in Chapter 7 where the massive gap between management rhetoric about the imperative to become a learning organization at the Rover car company and its actual practices in the succeeding decade bears witness to at best management inadequacy and at worst, deep managerial cynicism of a kind that typically accompanies inability to learn how to implement actions arising from new knowledge. Second, the kind of knowledge that is frequently high value nowadays is scientific, including social scientific. This is not new existentially but its scale and economic penetration are. Thus, so-called 'scientific management' was practised at Ford plants in the first quarter of last century, ultimately proving fatal to craft-based production methods in the car industry. Innovations from the Gilchrist-Thomas to the Bessemer processes and beyond in steel-making were scientifically knowledge-intensive, but new knowledge of electric arc production, for example, meant steel did not need to be produced mainly in ever-expanding works but in more localized, customized mini-mills where economies of scope (variety) could outweigh those of scale. This arose from the interaction of scientific knowledge about production and social scientific about management and markets.

A good example of an 'old economy' industry that has become more scientifically 'knowledge-embedded' is food production. It is shown in Chapter 6 how important agricultural research institutes in East Anglia continue to be to the development of agro-food businesses in the UK, not least in their questionable contribution to the application of biotechnology to this industry. In a more wide-ranging analysis of the embedding of scientific research in a specific food industry value-chain, Smith (2000) and colleagues mapped the nine key stages in the Norwegian chain and related these to their knowledge-content and knowledge suppliers. For preparation of raw materials, processing, preservation and packing thirteen different private and public laboratories were engaged. For hygiene and food safety, eleven, including some of those involved in preparation, etc., were involved, and for quality control, logistics, marketing and sales, eleven, again including some used in previous stages, were found to be knowledge suppliers. Thus the food industry in Norway and conceivably elsewhere in comparably developed economies is knowledge-intensive and relies on this characteristic to be competitive. But, as Smith points out, it is not a particularly research and development (R&D) intensive industry and its workforce is not in itself directly processing scientific knowledge, making as Castells (1996, p. 17) puts it: 'the action of knowledge upon knowledge itself as the main source of productivity'. The modern food industry is thus knowledge-using but not knowledge-creating, it learns but does not necessarily tutor scientifically and this must be one of the reasons why it is placed, and possibly misplaced, in the low-technology manufacturing category of the OECD (1999) index of 'knowledge-based industry'.

Thus, third, knowledge economies are not defined in terms of their use of scientific and technological knowledge, including their willingness to update knowledge and 'creatively forget' old knowledge through learning. Rather, they are characterized by exploitation of new knowledge in order to create more new knowledge. This need not be scientific or technological alone, it can be creative knowledge in the artistic, design or musical senses of knowledge. An example of this occurs with 'sampling' in music, which gives an innovative, creative role to knowledgeable deejays who, instead of merely plugging mass-produced records in linear fashion according to formulaic corporate interpretations of popular taste, deploy their own musicologies to evolve a new form from the imaginative appropriation of authentic sources, thereby creating a new authenticity. This is reminiscent of post-modern architecture, except that its realization in built form was not normally accompanied by the elevation of the property agent to star status in place of the architect. In the technological sphere, an example of knowledge acting upon itself as the main source of productivity is software engineering where written code forms the knowledge base for applications in the form of new code. Another is in biotechnology where the discovery of the genetic code structure allows 'sampling' or the recombination of DNA to produce therapeutic products for healthcare or food product application, while the de-coding of the human genome both creates opportunities for value-creation and opens up the need to discover the biochemistry of proteins, giving rise to the new knowledge field of proteomics. To the extent genomics and proteomics give rise to superior tests or drugs to those presently available at comparable cost, knowledge is acting on knowledge itself to enhance productivity.

In yet another version, the Digital Economy, the digitization of knowledge, meaning its transformation from analogue, real-world images, voice or text into digitized form on-line, on a CD-ROM or floppy disk means the initial form of the knowledge becomes a resource in a value chain. The next step after compression and storage of the digitized knowledge (as yet a resource to be mined, rather than knowledge having been acted upon to create new knowledge value) is for a knowledge-bearing user to access those elements in the digitized resource that they aim to transform into a new product. The entrepreneur will seek to make a profitable product, the organization may only seek to produce a socially useful product. The product could be a new media 'open learning' training course, a television programme or a cultural product, combining in a creative and imaginative way possibly dusty archive material into Internet or off-line content. The producer may contract to a marketing agent or publisher to sell the product and at each transactional point value accrues from knowledge acting upon knowledge. At specific points, such as that which demands a new or enhanced technology to 'mine' or locate the sought-after elements in the digital resource, innovative knowledge from, say, the software industry is brought into a conceivably high value adding but temporary (if the digital resource locator is itself commodified) position in the digital value chain (Williams, 2000). Customer and supplier may have found each other in Yellow Pages or, more likely, they may have been put together by a venture-manager with equity shares in both and who gains value from traded interdependencies. Far less likely, they might reciprocate without arm's-length exchange, at least in the customer exchanging their tacit knowledge of what service is required to mine the data 'warehouse', thereby giving the supplier the idea for a profitable innovation, in a relationship of untraded interdependencies. Either way, this is how clustering occurs in Knowledge Economies. The Digital Economy is an aspect of and electronic underpinning to Knowledge Economies, and they constitute what is often more popularly called the New Economy.

The imbalance problem and its governance

If clusters create imbalance and reinforce a predominant tendency present in market economies, what should be done to go beyond Malthus' question to develop some policy prescriptions for moderating the disparities they produce? This is really a different and much asked question of regional and industrial policies as well as those concerning development disparities at an international scale. Either the problem is insoluble, something the theoretical burden of the book presses ineluctably on the mind in Chapter 2, because it is an endemic feature of the generic mode of economic organization, or it is, if not soluble, capable of reversal in certain times and places. If there are cases of that, they can be investigated and lessons learned. They may not be directly applicable everywhere, but the notion that a policy accomplishment in one setting is not transferable to another both belies economic history and denies human ingenuity. How did Japan develop in the second part of the twentieth century? By copying the West. How did Japan industrialize in the second half of the nineteenth century? By copying the West. How did South Korea develop economically? By copying Japan. The economic history of every economy is littered with borrowings, some successful, many not, from other countries.

One of the problems of copying rather than learning and adapting with constant monitoring is revealed in the recent history of Japan. A thoughtful contribution by Nonaka and Reinmöller (1998) refers to the legacy of learning in the present Japanese downturn, which stemmed from a too obsessive copying of Western growth characteristics in mass-production industries. This required vast mobilization of national resources to stimulate consumer goods industries and the capital goods production to sustain them. But as hindsight shows, big investments in the West were being made, mainly by public investment in scientific research, that gave birth to industries now central to the New Economy, or the TMT sectors of telecoms, media and technology, where Japanese firms are not as strongly represented. The former is what Porter *et al.* (2000) call the uncompetitive half of the two economies of Japan. It is not so long ago that similar things were being said about the US economy's manufacturing weaknesses but less is heard about that presently (Dertouzos *et al.*, 1989). The key to overcoming the legacy of learning according to Nonaka and Reinmöller is recognition that no matter how great the efficiency and speed of exogenous learning, it is no substitute for endogenous knowledge creation. In Japan, the development strategy was insufficiently regionalized, too centrally managed and accordingly, too dependent on learning from other countries. Now policy needs to support regionalized knowledge creation to raise the diversity of possible innovations by stimulating inter-organizational interaction in networks and clusters. The case of Taiwan is presented as one in which such an approach was successfully pursued.

This is the central issue to be explored in this book, namely, to what extent can a decentralized industrial policy bring about regionalized industrial diversity by promoting networking and clustering and, by contrast, to what extent are such processes inaccessible to policy intervention? If the former is found to be feasible, in what ways have policy practices helped, and if the latter, are there specific points where policies are necessary even though the whole process is mostly market-driven? But to begin to answer such questions a good understanding is needed of reasons why cluster formation has become more pronounced, often characterizing Knowledge Economies, than for a century and what modern governance mechanisms need to be capable of if they are to have any role in moderating effects of imbalance while still assisting the processes in question. Nowadays, this needs to be understood particularly in the context of *regional* imbalance. Thus, in an increasingly information-saturated society and an advanced knowledge-based economy, it is fairly obvious that regions with universities have more potential to promote cluster-building activities than those that do not. Usefully, in terms of imbalance theory the university is one kind of innovation-supporting organization that is located in wealthy and poor, urban and rural locations. Hence, an equally obvious policy recommendation would be to ensure universities are present in all regions in a given country. But with a few exceptions, research shows universities to be less impressive than companies at stimulating fast-growth spin-out firms of the kind that produce functional clusters.

This is not an either/or position, as examples discussed in Chapter 7 show. But for the political process in many countries, cost-conscious decision-making has made too many governments trade off one option for another. In the university versus corporate trade-off, most governments would tend to shrink at the cost implications of investment in new universities but feel comfortable incentivizing firms in lagging areas to spin out new ones. In reality, a successful strategy can be shown to have both options rather than one or the other because start-up firms benefit from proximity to a knowledge centre that is familiar to their founders as an academic community, with all the networking opportunities and inherited social capital implied by that. Social capital is the extra value gained from interactions with familiar, trusted networks of acquaintances. But such firms also benefit from proximity to a customer from whom small commissions are vital in developing experience and a track record. Both large bodies will also benefit from interactions around research commercialization and the whole has the look and feel of a virtuous circle. A third dimension to add to the firm and academic aspects of this relationship is that of government itself as facilitator and financier of actions based on consensus. Regional governments may have influence over even rather expensive investments like universities and scientific research, but the big budgets for this kind of activity are usually national. National governments are good at setting frameworks for action but less so at detailed strategy in contexts with significant geographical variation, so here joining up government actions involves horizontal and vertical governmental relations, just as clustering does. This approach is usefully alluded to in the work on the 'triple helix' edited by Etkowitz and Leydesdorff (1997), but the regional governance and clustering dimensions are scarcely touched upon.

In the first two chapters of this book, the nature of economic imbalance is explored theoretically with a view to finding why it has been considered heterodox to place disequilibrium concerns at the forefront of economic development theory and to assess the relevance to the concerns of this book of those authors who have taken the heterodox approach. This means the work of Marshall (1916; 1919) plays an important role. As well as being a good industrial economist giving detailed accounts of latter-day clusters or 'industrial districts', he was one of those responsible for the marginalist revolution that took economic theory away from an understanding of disequilibrium by conveniently assuming the world of economic relationships operated in equilibrium. Nevertheless, it is the emphasis in his work on understanding the role of external economies, knowledge transfer, skills and learning among firms in geographically proximate settings that is of importance for our project from the outset. Marshall's belief in the market mechanism as the ultimate coordinator of highly complex inter-firm relationships is correct but it blinded him to the non-market exchanges that made them possible, and to which he referred as being 'in the air'.

This redolent phrase is of abiding fascination, as the neo-Marshallian research in Italy conducted by authors such as Becattini (1989) and Brusco (1989) demonstrates. It means people are talking freely about their business, that it is feasible for innovations to occur because, simultaneously, different entrepreneurs may deduce the same discovery from the collective humming of ideas and information, and anticipate the rest by getting a new product or service into production. It is a useful way of capturing the free goods quality of information that is immediately lost if a firm re-locates. It is fundamental to the supposed dynamic externalities capable of being creatively re-interpreted in conditions of proximity.

Yet there is also the nagging question whenever Marshall's notion of valuable knowledge being 'in the air' arises of just how much that was really valuable was in that ethereal condition. In those times, when technology was truly transformative in its effects on wealth and poverty, first governments of the day placed barriers on the export of machinery and knowledge, hence industrial espionage as well as emissaries were deployed by other countries to learn the secrets of the new forms of production. Moreover, within the clusters of the day it is probable that firms worked in complementary relationships that sometimes were a precursor to the formation of groups after acquisition, even if these may have been family-inspired initially. Accordingly, such techno-economic knowledge may not have circulated as freely as more generalized information. Finally, business associations were often the exclusive clubs of entrepreneurs in which innovations were formally and informally discussed prior to eventual proto-typing and production. But while principles would be discussed openly, patenting would protect against illegal knowledge application. In modern industrial districts in Italy it is argued that knowledge circulates freely because workers have to know their technologies thoroughly, and this must have been true to some degree in older ones elsewhere, hence the large amount of start-up activity typical of old and new clusters. But whether that extended to the capability to replicate from memory, drawings or, as occurred in Belgium, through the theft, over time, of enough mechanical parts to reconstruct a whole Yorkshire weaving machine remains open to question.

Schumpeter is clearer on this but not necessarily more accurate. His key contribution to the question of knowledge transfer was to introduce the notion of disruptive change or 'creative destruction' as the energizing process explaining capitalist development. An entrepreneur or R&D team of engineers makes a discovery and transforms it into a commercial innovation sold on the market. This is quickly followed by a swarm of imitators producing the same thing at less cost because they have been given the original idea and can reverse engineer it. This is what causes clustering in geographical space, although Schumpeter neither precisely wrote about such concrete spatial phenomena nor did he have much geographical sensibility generally, judged from his writing. More to the point, here is his discussion on monopoly and perfect competition. Both are limited in their real impact, the former because learning will occur, the latter because 'perfectly free entry into a *new* field may make it impossible to enter it at all'. Time and innovation erode monopoly while time prevents innovation 'perfectly promptly' being imitated (Schumpeter, 1975, pp. 104–5).

This gives us a most important clue about the reason why firms cluster. It is, on the one hand, to gain knowledge that can help them break monopoly through, on the other, seeking as near as humanly possible to gain something approximating perfect competition. The creative origin of the specific know-how being sought is spatially specific, so firms swarm around that geographical point. Capital is mobile, new knowledge is comparatively immobile. In knowledge economy contexts, in the age of the Internet, information is ubiquitous but knowledge is scarce. Thus, even if the human genome is put on the Internet, certainly this author would not and it is doubtful if many readers of this book would know what to do with it. Here lies the origin of economic and geographical imbalance.

The governance of this process is an extremely delicate matter. Landes (1998) argues that governments court danger when they intervene to effect institutional borrowing and when they try to force development by inducing change when institutions have not developed the required learning disposition or 'absorptive capacity' (Cohen and Levinthal, 1990). But it can be worse if governance bodies wait to be told by industry that they must do x, y or z to create the right conditions for innovation, clusters or entrepreneurship. Two examples are discussed in Chapters 3 and 7. The first concerns the way the European Commission was persuaded that it had an important potential role in developing a science and technology policy. This was done through the association of what, at the time, were Europe's 'Big Twelve' roundtable of technology firms, presenting an attractive vision of a new and powerful role for a body in search of distinctive functions that the states it represented had not pursued in this specific way. Approval was set against the perceived weakness of the 'Big Twelve' compared to American and Japanese firms at the time regarding technological innovation. Funds were earmarked for competitive bidding by cross-national business and research consortia for investment in applied technology projects that the Big Twelve were well positioned to win. In the two decades since then the innovation gap between the EU and its main competitors has not closed, except in mobile telephony, an industry that post-dates the establishment of the fund and is dominated by two states that were not members at the time. The second example can be briefly stated. For years Germany lagged behind the USA and the UK in biotechnology. This was of concern to the Federal Science Ministry that from the 1980s launched numerous programmes to promote commercialization of research it had also been funding in public and university laboratories. The ministry had no in-house researchers or special experts in the subject, relying for advice from large German pharmaceuticals companies. These had low absorptive capacity for biotechnology because their disciplinary origins lay outside the field. But they had an interest in accessing funding to promote their capabilities to use biotechnology and many of the grants they won from the ensuing contests went into investments in entrepreneurial biotechnology firms in the USA. When challenged about this state of affairs their response was that they would like to work with German firms but they were not as good as American ones.

'Rent seeking', a polite term economists use to describe economic practices of which these are both examples, is a kind of entrepreneurship but not one that is designed to meet the objectives of policies to correct economic imbalances. Both instances show the inappropriateness of managing innovation support programmes from a high governmental level where a natural business constituency is multinational firms with their own powerful agendas. Learning has occurred so that in both cases newer versions of policies adopt far more of a multi-level governance approach involving partnership between national and regional bodies. In this way, the specificities of local nuances can be incorporated much better. That is not to say that such policies are now immune from criticism and some are made in the relevant chapters. However, greater inclusivity of both the lower governance levels and the small and medium-sized enterprises (SME) sector are now features of both kinds of policy setting. One problem slowly and unevenly receding is that regional administrations are not everywhere as well organized and empowered as they are in federal settings or those where there is relatively strong regionalist political sentiment. Even in regions such as these, the promotion of innovative clusters has not been something which administrations traditionally had the competence to manage. That is changing, especially in the EU, where various innovation networking programmes at regional level have grown in take-up by regional governance bodies to over one hundred implemented regional projects by the turn of the present century. More regional authorities now have greater competence and confidence to conduct audits, build consensus and seek funding for actions to be implemented than hitherto.

This means the regions of the EU have developed some capabilities comparable to those practised by states and provinces in federal countries like Australia, Canada and the USA. In the latter, it is possible to have organizations conducting sophisticated knowledge-based industry audits, as the Massachusetts Technology Collaborative does, and state administrations committed to networking and cluster-building to enhance the innovative infrastructure by implementing policies based on such audits. Moreover, localized tax abatements and credits have been introduced to give incentives to investment in research and innovation. Planning deals have been done whereby individual firms can gain permission to develop knowledge-based industry in non-designated areas in exchange for offering free educational tours around the facility to science classes from local schools. Other planning issues are dealt with through consultation with municipalities so that those welcoming cluster developments receive new businesses and those wishing to conserve their character do not. This obviates the necessity for costly and conflictual decision-making that has slowed the process of establishing the infrastructure to enable rapid commercialization of the UK's global lead in some aspects of genomics in locations near to rural Cambridge. Issues of regional involvement in the complex governmental process of investing in and supporting science, technology and innovation are explored with the help of a multi-level governance approach in Chapter 3.

Social capital, trust and networks in learning economies

One thing concluded from research examined in discussing multi-level governance in the European Union is that regions showing greater mobilization and receptivity towards developing new forms of support for economic development were those with a well-developed sense of political identity. This could come from having clear, delegated powers that are comparable to those possessed by all regional administrations in a given country, as happens in federal states, or because of a strong sense of identity for socio-cultural reasons. In both types of case, capacity for lobbying superior power-centres is more pronounced as are levels of policy intelligence of relevance to developing innovative strategies or taking advantage of new opportunities. This links the argument of the book to a concept which bridges the spheres of governance and that of the economy and in Chapter 4 due attention is paid to *social capital* as a possible 'missing ingredient' from previous efforts to develop sustained economic development capabilities among firms at the regional and local levels.

Social capital is the expression of norms of reciprocity and trust between individuals and organizations that are embedded in a system of cooperation and favour exchange which gives advantage to those that belong, usually, to a particular locality or non-proximate community linked by ethnicity or religion. It has come to be analysed from an economic perspective as a consequence of advances made in studies of economic development in less favoured economies by authors such as Evans (1995) and Putnam (1993). Their insights are given space in a lengthy investigation of the role and limits of social capital in contexts of geographical proximity and distance. One key feature of the capability for cooperative action is that of collective learning. In the context of innovation in Knowledge Economies by firms and their support organizations, this phenomenon has been explored by authors such as Keeble and Wilkinson (1999). However, collective learning is seen by evolutionary theorists in economics and biology as a distinguishing feature of successful adaptation to externally induced change. It takes the form of a *selection mechanism*, an analytical device for explaining how one trajectory, which may be more successful than a previous one, or more successful than others have chosen is selected. To the extent that such selection mechanisms can be identified, they can play a powerful role in helping to manage change through policy or at least some form of collective decision-making.

The leading example of this is given in results by the evolutionary biologist, Allan Wilson (Wyles *et al.*, 1993) reported in de Geus (1997). The question being explored is, which species are the most evolved? Wilson developed a technique for measuring the evolutionary 'ticks' on the genetic clocks of species. Second after humans are birds, especially songbirds. Classic natural selection theory cannot explain this because it holds that evolution can only be inter-generational, yet birds had evolved faster than that. So something else must have accelerated their evolution. To explain this, Wilson advanced three selection mechanisms responsible for intra-generational learning: *innovation* or the capacity to invent and use new behaviour; *social propagation*, the capability directly to communicate skill from individual to community; and *mobility*, meaning collective movement rather than isolation and territoriality. The test for this thesis was the blue tit, the robin and the milk bottle.

In the UK, milk was traditionally delivered to the doorstep by the milkman. When this began, the bottles had no tops. Two bird species, blue tits and robins, learned to take the nutritious cream from the top of the milk. Both species prospered and even underwent some natural selection changes affecting digestion in consequence of this innovation. But the milk industry placed foil tops on milk bottles in the 1930s and stopped the 'free riding'. By the 1950s all the UK's blue tits were capable of piercing the foil tops but robins seldom learned how to do it and if one did, the knowledge never spread. Both species could innovate and communicate, but only blue tits had the networking or social propagation skills to spread the knowledge to the species as a whole. Blue tits flock and are numerous while robins are territorial and antagonistic to intruders and less numerous. Wilson's conclusions were that flocking constituted a process of collective learning.

Social capital is said to be cognate with flocking behaviour in that it is a characteristic of the group rather than any individual. It can be used but not appropriated individually and it cannot be acquired by accreditation only through membership of a community. In Putnam's (1993) research in Italy he concluded that social capital, measured by associative practices of communities in Italian regions, explained why the Mezzogiorno, with low associative organization, hence low social capital, was economically less accomplished and administratively less efficient than northern regions where both were high. These findings are examined critically in developing a clearer picture of the ways in which social capital and economic development may be interrelated. Conceptually this is assisted through investigating something of a shibboleth in this school of thought. This is Granovetter's (1985) idea of *embeddedness*, where he argued that for firms it is their personal ties and networks of relationships that differentiates them and their economic performance. He also added that strong ties can block development, but nevertheless held to the basic idea that embeddedness provides at least static externalities or comparatively costless community benefits. This is important but development is more comprehensively assisted by the exploitation of dynamic externalities that involve active, innovative engagements with local and non-local actors. Weak ties may involve 'tolerance of treachery' and that may be difficult in a tight-knit social setting. Hence, embeddedness is juxtaposed with *autonomy* and gains explanatory strength accordingly. A further 'conventional wisdom' to be exposed to scrutiny is the sociological notion that social capital cannot be appropriated. In the Knowledge Economy, venture capitalists have done just that, making those with the largest Rolodexes a magnet for New Economy businesses because of the social capital embodied in their lists of business contacts, and for which start-ups pay with equity shares in their futures.

Networks and clusters, old and new

In Chapter 5, the book moves towards its more practical concerns with actual forms of social capital, learning and cooperative, trust-based advantage by reference first and briefly to its discovery in an intra-organizational sense. This is best represented by reference, not for the first time, to the work of de Geus (1997), mentor of the learning organization guru Peter Senge (1995). The former was a senior executive in Shell charged with future studies and planning to meet a range of sometimes almost unthinkable scenarios such as what would Shell do if oil disappeared. In the process, he became a strong proponent of a corporate learning disposition for firms of all kinds. One project of his team was to find the oldest company in the world. This turned out to be a twelfth-century Japanese metal manufacturer that still exists as part of the Sumitomo Corporation. In this case, as with other firms that have displayed longevity, they had done so by changing their business focus every twenty years or so. Thus, gunpowder manufacturer Du Pont became a chemicals and now paints and polymers firm, rather like ICI who spun off its pharmaceuticals arm as Zeneca in the mid-1990s.

Nokia began as a forest products company, then paper and pulp machinery became the main line produced, then rubber, cabling, computers, data services and now it is the leading supplier of mobile telephony to the world. Nokia's contemporary exploitation of social capital in Finland is explored further in Chapter 7.

However, when the system into which a learning and change disposition is to be integrated is an externalized inter-organizational one rather than a corporation, the task is more difficult. Some attention is therefore devoted first to US practices in building 'economic communities'. Henton *et al.* (1997) argue for strong leadership to maximize, for profit, social capital effects and cite cases like the latest revival in Silicon Valley's fortunes through the association called Joint Venture Silicon Valley, the revival of Cleveland and the transformation of Austin, Texas. Austin's evolution from sleepy campus and government town to one of America's leading New Economy clusters is presented, as are the others, as cases of learning clusters based on cooperative advantage. The *New York Times* eulogized the place in January 1999 as:

> at once the least Texan and the most Texan of cities, with a burgeoning hi-tech industry, a University population of over 50,000, the endless carnival of Texas statehouse politics, and a music and restaurant scene that would be envied by a city twice Austin's size. Austin is one of those cities like Seattle and Santa Fe that gets so much praise you wish you could hate it.

Economic growth was running at 9–10 per cent annually in the 1990s and 30,000 new jobs were being added each year, some 200 start-up technology companies were founded there each year and it had a relatively recently established specialist business support system, largely private in origin. This was propelled forward by the actions of community entrepreneurs, particularly George Kozmetsky, founder of the Austin economic model, known subsequently as the 'triple helix' of government, industry and university interaction (Etkowitz and Leydesdorff, 1997). This led to a successful cluster policy in which all three actors connected also to entrepreneurs, sources of technology, venture capital and an innovation infrastructure of lawyers, accountants and business incubators. The policy key is leadership on each part of a strategy to retain and attract existing business and to grow new ones, and each leader is drawn from the 'economic community'. Luck plays a part, but even bad luck can be parlayed into future benefits. Thus the University of Texas at Austin is the second richest after Harvard in the USA, hence the world. But it started with a handicap in that it could not be a federal land grant university as Texas contained no federal land in the 1870s when it was established. Instead it received 5,000 acres of cotton and forest property, plus building plots. So successful was the cotton crop in the first year that the South Texas legislature re-appropriated the land and gave the university in perpetuity 2.2 million acres of wasteland in exchange. Not a very auspicious start, except that beneath the wasteland were discovered enormous resources of oil. Now the University of Texas' endowment stands at $7 billion. Some of this windfall was used to attract 'anchor' projects like the semiconductor research consortia Sematech and MCC, which gave Austin a local high technology edge. This attracted 3M, Dell,

IBM and Motorola to set up applied research facilities. Dell's just-in-time production system means that, as well as its 20,000 Austin employees, it sustains an equivalent number in its co-located supply chain. But New Economy clusters are not immunized against economic downturns as Dell's cut of 1,500 jobs in 2000, and expectations of lower rates of city job growth 2000–2002, down to 3.7 per cent from 5.2 per cent signify.

Networking of this market-led kind contrasts with many received examples of networked regions and localities discussed in detail in Chapter 5. The Italian industrial districts, some of which are entrepreneurship-driven with public involvement well in the background, most notably in north-eastern regions like Friuli and Veneto, are a little like Austin, with Treviso, where Benetton started, being the most well known. In Pordenone, Veneto is a remarkable cluster of engineering firms specializing in hydrological equipment ranging from horticultural irrigation systems to hot-tubs. Indeed, nearby is the birthplace of the eponymous Mr Jacuzzi who made his fortune when he emigrated to Arizona. Only the Chamber of Commerce and the business associations are of real consequence to the collective development trajectories of these firms since they supply the services for export and insurance, marketing and licensing that such firms find carry transaction costs that are irksome to bear. In the more celebrated Emilia-Romagna there is a more interventionist regional government that works with the economic community supplying needed services more adroitly than the market. Being traditionally a fiefdom of the Italian Communist Party made this an intriguing apparent exception to the rule observable in the Soviet system and elsewhere that 'the left can't manage'. For in all spheres of governance within its jurisdiction the region was outstanding; it was even the first to respond to Putnam's (1993) research test to ask each Italian region to send him some promotional material, unlike Calabria, whose response is rumoured to be still awaited.

Emilia-Romagna attracted US business consultants of a heterodox persuasion who sold a policy message that cooperative small firm systems should no longer be thought anachronistic but rather, in line with the argument of MIT academics Piore and Sabel (1984), as superior, more competitive and less alienating work places than large mass-production corporations. Whether or not this was entirely true, it proved attractive to meso-level governments that had some purchase on small firms policy and to smaller left-of-centre countries with a preponderance of small and medium-sized enterprises, like Denmark. In Chapter 5 there is an extended account of Denmark's experiment with building inter-firm social capital for profit through its network programme. This was as influential in the 1980s as clustering became in the succeeding decade. The two concepts are complementary but distinctive, and space is devoted to demonstrating conceptually and practically how they differ. The chapter concludes with a full exposition of how these policy forms realize and give expression to many of the ideas discussed in the foregoing chapters like imbalance, localized cooperative trust-building, learning and social capital. It also, crucially, asks whether policies designed to build networked economies succeeded or not and learns in the process some new lessons about fundamental flaws in policy-design and evaluation.

The next two chapters focus entirely on clusters. These are seen as the state of the art of industry organization at the start of the new millennium and a substantial amount of space is devoted to analysing a range of different types. This exposition begins towards the end of Chapter 5 but by Chapter 6 more traditional clusters in older and newer industries have been explored, and the interest turns to a critical investigation of what emerges as a fairly generic feature of the TMT or New Economy industries, which is that they are all found to operate as Knowledge Economies, knowledge acting upon knowledge for productive gain, in clusters. Hence Chapter 6 investigates the extent to which a New Economy exists, what various authors have said about its salient features and whether, as a consequence, there is really anything new about the New Economy. The interim conclusion is that there are a few distinctive features about TMT that warrant a distinctive designation and that these are not necessarily to do with clusters but reward systems. Most of what are thought to be New Economy features look remarkably like characteristics of Schumpeterian swarming at any point in the history of capitalism, robber barons and anti-trust suits included. The proposed break-up of Bill Gates' Microsoft was resonant of the dismemberment of Rockefeller's Standard Oil and the rapaciousness and hard-nosed management of Intel under Andy Grove could be modelled on that of Carnegie in steel and Huntington or the other railway companies.

Three kinds of motive force are propelling clusters of the New Economy type. Curiously, large-scale public-sector funding is fuelling the process in two of the three kinds. Military expenditure is the key source for triggering the evolution of information communication technologies (ICT) clusters such as that found in Silicon Valley, as has long been known (Saxenian, 1994), and even though defence budget cuts mean it is not the main motive force it once was, along with public space programme funding, it is still present and important. Dot.com and software start-ups arise from an evolved ecology of firms that profited from rocket science and its baroque arsenal. Rather than conduct yet another investigation of the most recent configurations in Silicon Valley, Chapter 6 chooses as its exemplar of the booming military equipment cluster that has now developed more benign civilian research applications, a Texan case that is not Austin but Richardson, home to the Telecom Corridor. This is a northern suburb of Dallas employing 78,000, of which well over half work in telecommunications firms like Nortel, Ericsson, Alcatel, Fujitsu, Samsung and Texas Instruments, and their suppliers, among which are hundreds of corporate and university-originated start-up businesses. The cluster character of the area is underpinned by associative organizations and collaboration on cluster issues like pre-competitive research and human capital provision. Lobbying by industry has ensured Richardson is a global telecoms hub with access to the scarcest, leading-edge infrastructural installed-base, the Internet 'backbone'.

The second type of publicly-funded cluster is typified by biotechnology as conducted in Massachusetts and, in the UK, the south-east of England. Enormous public research budgets are directed towards basic scientific work in the life sciences. In the UK it is of the order of £1 billion per year including charitable funding from the likes of the Wellcome Trust. In Boston alone something of the order of $770 million in public research grants passed through the biotechnology cluster in 1999. Naturally, the glamour is reserved for the dynamic start-ups or, by now sometimes relatively mature drug discovery firms, and perhaps the pharmaceutical giants that supply milestone payments to such firms as they go through the laborious process of moving from 'proof of principle' through multiple rounds of trialling until they may at the end of ten years have a product that 'big pharma' can market and distribute. The Cambridge (US) and (UK) clusters are exemplary cases of fully-fledged technology clustering with strong science bases, exploitation organizations, venture capitalists, specialist patent and IPO lawyers and regional industry associations that lobby government and provide collective services. Accounts of the key lineaments and characteristics of both are presented in Chapter 6.

Then, attention turns to a third type of cluster, which can be broadly defined as 'entertainment' but is referred to as new media in which large public budgets are absent but start-ups and large corporations cluster in distinctive downtown areas like Soho in London and SoHo in New York. The boundaries between industries blur considerably in this core Digital Economy sector, more so than normal, although cross-sectoral blurring is a distinctive characteristic of clusters. Thus entertainment involving new media ranges from cinema to hand-held computer games equipment and the animation, computer graphics and other design skills that routinely make up the products in question. These are creative and innovative activities that may have formalized higher education in the network, but are not as strongly tied in as in the other two kinds of cluster just described. Advertising agencies may perform the function of 'basic research' or it may be the presence of a large network TV broadcaster. Or it may be just what's going down on the street that is the source of stimulus. Whatever the nub of the cluster, it attracts small start-up businesses to occupy incubators or sub-sub-leased space in the funkier parts of town and finds a burgeoning market accordingly.

Chapter 7 has two main aims. After investigating the sources, practices and relationships characteristic of these creative, innovative, learning clusters, attention is shifted to understanding the mechanisms at play in bringing about the demise of clusters. Here evidence from three UK clusters in various stages of decline as clusters is presented. The furthest gone as a cluster is that of the silk industry in Macclesfield, once one of Europe's four leading silk-producing centres, it has attenuated to a vestigial condition, unlike its traditional rival of Como in Italy, an industrial district that retains and augments its economic vibrancy. Macclesfield, by contrast, while prosperous, displays its silk traditions as part of the UK's heritage industry of museums and replica work-spaces of a bygone age. Its fate was nearly shared by a second instance, the ceramics cluster of Stoke-on-Trent familiarly known as The Potteries. This was still a £300 million net export surplus industry for the UK in 1999, but its firm ownership became concentrated into a duopoly with cross-shareholding and all its traditional SMEs have disappeared or been merged into the duopoly. There is, at the eleventh hour a multi-level associative effort to revitalize the district by stimulating new firm formation and recreating the smaller firm profile that typified The Potteries in its heyday and which continues to typify successful ceramics clusters elsewhere in Europe from Sassuolo in Italy to Castellon in Spain and Coimbra and Aveiro in Portugal.

Second, Chapter 7 asks whether clusters can be planned or built by examining instances from northern Europe where precisely that task has been pursued. The initial focus is upon the Nordic countries where, in Finland particularly, some ten clusters of high technology industry each housing as many as two hundred endogenously generated firms are found. In each case the heart of the Knowledge Economy cluster is a university and present also is a branch of a larger firm like Nokia. This also happens to a considerable extent in Sweden and there are success stories in Gothenburg, Lund and Linköping. So one way of successfully building Knowledge Economy clusters, in some cases in unpropitious circumstances in regions of older, declining industry like Gothenburg, is to design them around universities and provide appropriate subsidies.

The other example chosen to illustrate cluster-building strategy with multi-level governance of policy and funding is Germany. Since 1995 the BioRegio programme has been implemented to build biotechnology clusters. The chapter provides a detailed account of the mechanisms involved and an interim assessment of the likely success of the programme. The overall conclusion is that carefully designed cluster-building policies can work but that it is extremely difficult, though not impossible, to do so from scratch.

To recap, therefore, Chapter 1 has set the scene with some key definitions and a taste of some of the material explored in this book. Chapter 2 is concerned with Knowledge Economy clustering as a modern version of an abiding feature of capitalist economies, namely, their imbalance, a source of government policy which has sought to moderate the inequalities entailed, since the Great Depression of the 1930s. The Knowledge Economy is much more uneven in spatial terms than the economy it is beginning to supersede and governments everywhere need to pay attention not only to digital divides in Internet consumption but the even starker imbalances in the locations of Knowledge Economy production. Chapter 3 engages with the complexities of governance and modes of intervention in a world which is increasingly managed through rules and processes of negotiation involving multi-level governance. Chapter 4 is an in-depth exploration of the roles of learning, trust and social capital, thought by some to be the basis for engagement in production as well as consumption of the fruits of Knowledge Economies, found universally in clusters but by no means ubiquitously for reasons of imbalance. Chapter 5 is about networks and clusters, a definitional voyage that is exemplified by efforts by policy-makers to induce social capital for profit by building both, sometimes successfully, sometimes not. Chapter 6 extends the analysis by focusing hard on three distinctive types of cluster-based Knowledge Economy and explains how they grew in hitherto unclustered settings. Chapter 7 looks partly at clusters that failed and policies that are in place to build clustered Knowledge Economies, often centred upon university research laboratories and the commercialization of science. Finally, Chapter 8 summarizes the book and points to the future key features that Knowledge Economies will tend to thrive around, suggesting a more nuanced role for public initiative and a greater requirement for learning from successful aspects of private Knowledge Economy building based on cluster support as a profit-earning activity.

2 An evolutionary approach to learning, clusters and economic development
Theoretical issues

Introduction

This chapter presents an abbreviated history and analysis of major themes in economic development theory. Its aim is to argue that such thinking took a turn away from realistic concerns with explaining growth and change in the space economy that were obvious in or could be inferred from the work of the founding fathers. As in rest of economics, the power games of scientific legitimacy led regional and developmental economists down the path of mechanical analogy, marginalism and equilibrium assumptions. Only a few outsiders retained an interest in understanding economic development as an institutional–cultural process characterized by social interaction, even cooperation, among humans rather than rationalizing automata, and sought to explain disequilibria in historical and geographical terms.

These issues are explored in the three main sections of the chapter. The first contrasts the work of theorists adopting an equilibrium as against a disequilibrium stance and concludes that much the most interesting questions arise from the disequilibrium thinkers. The second section deals with the response from the neoclassical equilibrium wing to the abiding problems posed for their science by disequilibrium. Among the most interesting but still unreal contributions here are those of Krugman (1991; 1995), prominent among 'new growth theorists'. The final section explores evolutionary economics and draws out four strands that contribute to the key features of that perspective. This section argues that the kind of concerns and questions raised in evolutionary economics are more appropriate for both synthesizing equilibrium and disequilibrium analyses and producing insights and understandings of the most direct relevance to practical questions of unbalanced economic development confronting us today.

The classical and equilibrium approach

Growth and change in economies have been central objects of investigation for development economists since the dawning of the classical era with the work of writers like Adam Smith and David Ricardo. For the former (Smith, [1776] 1976) a key factor was the productivity improvement that followed the establishment of a division of labour in the production process. Whereas craft production entailed single workers conducting a variety of tasks to produce goods, thereby mastering many different skills, industrial production – as in his famous case of the production of pins or needles – involved separating the manufacturing process into a number of phases: cutting wire, honing it, inserting the eye, polishing and packaging. These tasks could then be allocated to workers who would then become specialists in a single phase of production, handing the semi-finished product on to those responsible for the next phase until the packaged final product could be shipped. Enormous increases in productivity and wealth accruing to owners ensued from this discovery.

Linking this with one of Ricardo's (1817) key critiques of Smith, that the value of a commodity depends not on the price it can fetch in the market but rather on 'the relative quantity of labour which is necessary for its production' (ibid., p. 11), led to the theoretical insight which has proven something of a touchstone for development theorists ever since: comparative advantage. Ricardo's theory stated that differences in labour productivity explained the development of trade between economies. More fully elaborated later by Heckscher and Ohlin (see, for example, Ohlin, 1933), the comparative advantage thesis is that economies differ in resource and other production endowments. Where they possess specific resources, including labour, in abundance they will specialize in the production of goods embodying that endowment.

For *regional* development theorization, the classical insights were of fundamental importance, though the work of Weber (1928) and Lösch (1952) became the cornerstones of future work by regional analysts of the neoclassical persuasion, particularly Isard (1956) and his disciples. Classical thinking assumed economic activity would concentrate where endowments of particular value were to be found. Where the costs of exploiting natural resources were cheapest, resource-using firms would congregate and labour would migrate to such regions. Where labour was the required resource, otherwise unconstrained firms would locate where it was found in abundance. For Weber, location of firms was influenced by natural resources in the following ways: where substantial weight loss occurred in production, location near the source was optimal; where substantial weight gain occurred, location near the market was to be preferred as the rational option.

Recognizing that the locational influence of transport costs was always likely to be less than that of labour, Weber went on, following Marshall (1919), to examine the centrality of *agglomeration* by firms in a specific region or sub-area as a motive force of economic development. External economies accrued from the geographical association of firms seeking complementary inputs from others. Whereas in early stages of industrialization a resource-exploiting industry such as steel-making would be liable to make rather than buy its own capital equipment, over time, such production becomes too specialized as technology improves and companies shift from internal self-sufficiency to externalized acquisition of equipment on the market. In Stigler's (1951) analysis of this process, it is argued that, as the market for such specialized inputs grows, sub-sections of the parent company will likely be hived off as separate firms which, for financial, cultural or basic market reasons, will tend to co-locate near the customer-parent, thus setting in motion the spatial agglomeration or clustering process.

The proximity effect of divestiture was not seen as an iron law by Weber, rather, a dominant tendency where complementary industries found advantage in sharing facilities, know-how or access to common services in a specific region. Where these were outweighed by some other economic advantage, such as the need to access particular workforce skills, or more valuable markets, that would draw some away from their location of origin. Nevertheless, for Marshall a key task was: 'to examine those very important external economies which can often be secured by the concentration of many small businesses of a similar character *in particular localities*: or, as is commonly said, by the localization of industry' (1962, p. 221). An interesting question which arises from this is precisely, what is meant by businesses 'of a similar character'. This is because, at the time Marshall was observing industrial agglomeration, firms were often, though by no means exclusively, what might be called imitative spin-offs, i.e. clones of an original tinplate firm in Llanelli co-locating in nearby villages or a Sheffield cutlery firm spinning off laterally into a nearby workshop or Lancashire cotton spinners into Oldham, each of which Marshall wrote about.

This process still occurs in contemporary Portugal, for example, or in Attiki in Greece, economies that show high rates of localized horizontal disintegration of small metal manufacturers, furniture and textiles firms in north-central Portugal and clothing firms in Athens (Cooke and da Rosa Pires, 1985; Halaris *et al.*, 1991; Cooke, 1993; de Castro *et al.*, 1996). But in terms of the derivation of capital goods inputs, such firms source from abroad, hence the productivity and enhanced value-added gains that Weber and Stigler talk about are not yet localized, except insofar as marginal improvements accrue from these piecemeal modernization processes. Of greater moment is the elaboration of the division of labour upstream and downstream into specialized machinery and advanced services production, which is found in contemporary Italian industrial districts such as Prato and Carpi for textiles and Sassuolo for ceramics (Russo, 1985; Cooke, 1996; Dei Ottati, 1996).

The former – imitative – kind of externalization produces localized agglomeration effects based on relatively low learning and innovation gains, not least because businesses tend to remain locally competitive, retentive of information and mistrustful of their peers. In the other, vertically disintegrated, kind of externalization there are greater opportunities and requirements for interaction based on business complementarities rather than head-to-head competition. Learning effects accrue from user–producer communication and innovation from the exchange of tacit knowledge, as compared to reliance on more codified knowledge available in manuals and through straightforward customer training in the use of new machinery developed elsewhere. In Marshall's (1916) *Principles* the advantages of agglomeration to businesses of a similar character were seen as arising from the existence of localized pools of appropriately skilled labour, the opportunities for developing specialized product and service inputs, and the availability of external economies from technological innovation.

In fact, the industrial district Marshall defined includes both strong horizontal and vertical specialization between firms, not least because in his view the market mechanism of arm's-length exchange was the primary motor of the agglomeration process. This point is well made by Robertson and Langlois (1995) who interpret Marshall in this way as follows:

> Firms tend to be small and to focus on a single function in the production chain. Suppliers of intermediate goods commonly sell their stocks locally, within the district, although the final products may be marketed internationally. Firms located in industrial districts are also highly competitive in the neoclassical sense, and in many cases there is little product differentiation. The major advantages of Marshallian industrial districts therefore arise from simple propinquity of firms, which allows easier recruitment of skilled labour and rapid exchanges of commercial and technical information through informal channels. As Marshall described them, industrial districts illustrate competitive capitalism at its most efficient, with transaction costs reduced to a practical minimum; but they are feasible only when economies of scale are limited.
>
> (Robertson and Langlois, 1995, pp. 548–9)

This means, as they interpret Marshall, advantage accrued from local economies of scope, whereby firms might produce a range of more or less customized intermediate goods for local customers, and where economies of scale were also localized to the limited production of identical inputs for local customers. This combination of localized scope economies and limited scale economies nevertheless allowed the final producers, those in touch with the market, to reap for the district non-limited scale economies of the kind normally associated with large, stand-alone, vertically integrated firms.

To return to Weber momentarily, he, unlike Marshall, saw technological progress and innovation as being likely to cause dispersal of business activity away from the agglomeration either because technology was externally derived and thereby an attraction factor in its own right or because technology changed the weight of products and therefore transportation costs, encouraging a locational shift towards the market. As Weber put it, 'everywhere the mechanisation of production creates new stages of production which have *independent* locations' (1928, p. 193). Though this was not a major concern of Weber, it became an important lacuna in the work of later writers such as Lösch (1952) and Isard (1956), as it did with neoclassical economists more generally. Technical change eventually became treated simply as an exogenous variable to be purchased at arm's length, as were most other required inputs to production.

But the larger problem for explanations of regional economic development processes following Lösch's critique of Weber, and its subsequent elaboration by Isard and others, was the abstractionism they practised as they sought mathematical precision in the pursuit of 'regional science'. Strong assumptions of perfect competition, perfect access to resources, perfect information and spatially undifferentiated endowments of technical knowledge, consumer preferences and market areas were introduced in the pursuit of a general equilibrium model of the economics of location. Thus from a promising beginning in which attempts were made to explain, in realistic ways, the functioning of space economies, the subject moved inexorably into a discipline seeking to model the space economy. In other words, reality was to become merely a special case (for a fuller critique, see Cooke, 1983).

The disequilibrium or Schumpeterian approach

One writer who did not follow this particular economics pathway was Schumpeter and it is his work, developed by neo-Schumpeterian theorists of innovation, in particular, that leads us obliquely back to more realistic theorization of economic evolution. Schumpeter shared with Marx a view of the capitalist economy as a system characterized by evolutionary turbulence with a self-engendered capacity for '*creative destruction*' derived from the incessant pursuit by its progenitors of technological and organizational innovation. Thus, unlike any of the writers discussed so far, with the exception of Marshall for whom it was an incidental, Schumpeter places technical change at the endogenous heart of his theorization of economic development.

In his earlier writing (Schumpeter, 1912, trans. 1934), partly influenced by Nietzschean philosophy and in admiration of American entrepreneurship, he identified as the key agents of change exceptional individual entrepreneurs, willing to risk all in an act of will to innovate despite the barriers of the present and the uncertainties of the future. He even spoke rather poignantly of the fate of the Austrian innovator who, at the turn of the nineteenth century, often innovated as first mover only to be caught and overtaken by a second, more numerous group of imitators adapting to the new realities more quickly (Freeman, 1994). In a critique of this aspect of the work of the early Schumpeter, Malmberg and Maskell (1997) suggest this meant that, to the extent a spatial dimension can be discerned in this stage of Schumpeter's work, it is one in which firms agglomerate as imitative competitors, rather than innovators in their own right. However, that is probably too restrictive a reading since, as we shall see, imitation is not straightforward, and entrepreneurship is normally understood to rest on at least some competence to spot an unfilled niche in the market. The key Schumpeterian concept to deploy in relation to spatial swarming is that of incremental innovation stimulated by an initial radical innovation. This chimes well with Zucker *et al.*'s (1998) notion of the importance of 'stars' as an agglomeration factor.

However, a key distinction is made, especially by neo-Schumpeterians such as Freeman (1992; 1994) between radical and incremental innovations. In Schumpeter's terms, the 'heroic' entrepreneur is responsible primarily for radical innovation and the imitators are responsible for the incremental ones that follow. Those that failed to adapt to the new circumstances also suffered 'the fate of the Austrian entrepreneur' and went out of business. But there is a tension between opportunity and capacity to act upon an innovation. In an example discussed by Andersen (1992), Schumpeter pointed to the real case of a nineteenth-century US railroad company that opened up a region, built grain elevators, put in infrastructure for farmers and even produced manuals on agricultural methods in order to fill the gap or moderate the tension between innovation and take-up. This clearly shows a heroic, pioneering entrepreneurialism on the part of the innovating company, but also a crucial dependence for success on the willingness of others to interact, imitate and even innovate incrementally. Hence some degree of cooperation and learning are necessary for competitive risk-taking to prove rewarding.

The essential point here is that imitation is more difficult than Schumpeter allows. First, asset-specificity, or the embodied know-how of the innovator, gives that agent a competitive edge. The successful innovator will reinvest in the future know-how that will keep that edge. Second, tacit knowledge takes time to communicate and understand and has costs attached to it. These are shown, also by Zucker and colleagues to have a strong dependence on formalized and time-limited interactions rather than being like free goods. Third, innovators will have their own linkages with other firms or sources of the knowledge that give them their initial asset specificity. These forms of exclusivity reinforce the difficulty of learning and innovation in many circumstances. Despite these reasonable caveats, there are numerous well-cited examples where successful imitation has taken place, from the copying of British textile-production machinery by Belgian and German firms in the nineteenth century (Keck, 1993) to contemporary 'cloning' by Asian firms of Western personal computer technology. Hence, critique of Schumpeterian assumptions of the ease of imitation only applies to successful resistance to imitation by firms, especially those in geographical clusters, who get a breathing space by calling in favours, or externalized assets, where they operate in a cooperative, trustful environment rather than an arm's-length exchange one. The fate of the Austrian entrepreneur may have been that he was, in Schumpeterian terms, too Nietzschean, i.e. too heroic.

The later Schumpeter came to the view that as capital had become concentrated in large, oligopolistic corporations, the wellsprings of innovation would move from the individual entrepreneur to the research and development (R&D) laboratories of these organizations. Freeman (1994) points out that Schumpeter even welcomed this development, marking as it did a change in his view of the importance of individualistic competition to economic dynamism towards one in which monopoly had a role to play subject to regulation. He even equated the R&D engineer with the heroic entrepreneur of yesteryear. Schumpeter thus constructed elements of the base from which the neoclassical hegemony over economics would be challenged by evolutionary economic concepts in which learning, innovation, entrepreneurship, imitation and economic change and development play important roles. However, in the hands of one theorist of spatial imbalance, Perroux (1950) who drew inspiration from Schumpeterian theory, a decidedly less than evolutionary outcome ensued. Perroux's adaptation of Schumpeterian thought on learning and innovation became transmuted into the growth pole concept. This in turn was applied as intended by regional planners in France where it also mutated into the technopole idea, subsequently to be taken up by the Japanese Technopolis programme. The policy outcomes in each case have proved largely disappointing (Castells and Hall, 1994). Why should this be?

In a study of this question Andersen (1992) produced an interesting answer. He showed that the origin of growth pole theory lay in Schumpeter's ([1928] 1951) discussion of the direct and indirect effects of radical innovation. These create the context for the adaptive decisions which may or may not result in 'the fate of the Austrian entrepreneur'; where they do not, then the imitative adaptation and incremental innovation occurring around the radical one constitute 'clusters of innovation', the 'swarming' process that, Schumpeter argued, set off bursts of dynamic economic growth. Given a propellant industry and adaptive, incremental innovators it was but a short step for Perroux to imagine the idea of a regional growth pole capable of inducing development in a backward regional economy. It is the motor element for an innovation system as Andersen perceived. Perroux started to go wrong when he began to operationalize the concept in ignorance of the time compression diseconomies involved in localized learning along with the various specialization assets possessed by the radical innovator. Perroux simply stressed the importance of co-location or simple spatial agglomeration. In a sense, he proposed short-circuiting the evolutionary process by substituting space for time. To achieve this, Perroux advocated, as did Isard, that the then new and relatively untested technique of input–output modelling be applied to given industries. The policy recommendation was then to co-locate the customers and suppliers with the strongest linkages in the model outputs. Input–output analysis is a useful, often static or snapshot, descriptive tool but even when dynamized it can *explain* nothing about the nature, institutional setting and vintage of the relationships it identifies.

As Andersen (1992) puts it: 'The policy prescription for development was then to invest in important core industries which for one reason or another were not present in the nation under consideration' (1992, p. 70). The result was expected to be that this exogenous propellant would cause the creative destruction or renewal of the adaptive parts of the economy bringing the incremental innovators into the linkage patterns or supply-chain of the growth pole core. But because the model was based on an unusually mature industrial system – Perroux's actual model was the German Ruhr complex of coal, iron and steel, chemicals and power engineering industries – and time, along with community, culture, politics and institutions, was excluded from the equations, the application of growth pole policy failed. It is true that Perroux warned against this transformation of economic decision-making and maturation into geographical determinism, i.e. the belief that simple co-location by agglomeration would cause 'synergies' between business actors, but policy-makers and especially development-thirsty political actors paid no heed. As Andersen concludes:

> this translation of Perroux's argument is, unfortunately, radically wrong. The tight 'linking' of industries revealed by the input–output tables of the most advanced countries has no necessary connection to growth poles. On the contrary, it probably indicates a 'mature' situation with routine deliveries and few possibilities of change and development.
>
> (1992, p. 71)

As much as anything, such radical disturbances to a given space economy constituted the reason for failure precisely because of their radically innovatory character. This non-evolutionary approach, less conscious of the need for social processes of interaction-stimulation and learning than even Schumpeter's exemplar railroad was, produced in most cases confusion and disappointment, most glaringly at Gioia Tauro in Calabria, southern Italy where a vast, isolated port facility for decades bore witness to the folly of the 'quick fix'.

Disequilibrium and evolutionary development

Moving on from Schumpeter, this chapter retains its explanatory intent towards disequilibrium by reference to the work of more recent theorists compatible with a broadly neo-Schumpeterian framework, which is also germane to understanding imbalance in regional development. This becomes especially pertinent in assessing the importance to learning economies such as clusters of the dynamism of contemporary development processes in situations where learning is an embedded feature of the institutional structure. Bridging a summary of the inadequacies of neoclassical economics and attempts to salvage its most treasured propositions, to our exposition of the evolutionary economics approach is the work of Gunnar Myrdal and Albert Hirschman.

Having made the link from Schumpeter to Perroux, the next step is to consider the approaches of Myrdal and Hirschman, also well-known theorists of disequilibrium. Schumpeter's focus was on the turbulence, 'creative destruction', and cyclical fluctuations of dynamic economies. His theory of entrepreneurship and the unbalanced impact of radical innovation was founded on disequilibrating rather than equilibrating effects, a factor distancing him from the neoclassicals. Perroux's growth pole concept is a profoundly disequilibrating mechanism when applied in a regional context, especially when it works, as, to some extent and for a limited period it did at Fos-sur-Mer near Marseilles where it was centred upon a now-defunct steel plant. The point of growth poles, as Malecki (1991) notes, was their imputed ability to justify unbalanced growth policies to induce development in persistently lagging regions.

Both Myrdal and Hirschman shared Perroux's enthusiasm for the propulsive effect of a core industry combined with localized linkages to supplier firms, and both spoke positively about, respectively, the spread and 'trickle-down' effects that could be anticipated as growing incomes and output radiated outwards into the regional population from the growth centre. The diffusionist imprint of the times is clearly visible in the framing of these ideas, but so to a considerable extent is the temporal, evolutionary dimension in that immediate trickle-down effects were not necessarily assumed. More importantly, though, both authors theorized the negative aspects of economic growth as well as the positive. Myrdal wrote of 'backwash' and Hirschman 'polarization' effects of growth centres drawing people and investment from the periphery to the centre. Indeed, both the positive and negative effects were conceived as operating simultaneously, with peripheral growth – again, time-dependent – a question of whether and when 'trickle-down' or 'spread' effects came to dominate 'back-wash' or 'polarization' effects.

Myrdal, in particular, advanced a theory of development (or underdevelopment) based on the idea of 'cumulative causation', the cornerstone of his disequilibrium perspective. Market competition produces imbalances in the use of regional resources, tending to increase rather than decrease regional inequality. Profit earned in a regional industry will be invested in the location carrying the highest rate of return which, if found in a different regional industry, or more likely, a different region, will cause the accumulation of growth capacity in the more attractive industry or region. Thus the attractions of one region will tend to be reinforced, as will the detractions of another. As Myrdal put it:

> this is so because the variables are so interlocked in circular causation that a change in one induces the others to change in such a way that these secondary changes support the first change, with similar tertiary effects upon the first variable affected, and so on.
>
> (1957, pp. 26–7)

The interest in this quotation lies in the evolutionary tone of the description of circular or cumulative causation which recurs in more modern guise in evolutionary innovation theory as the 'path dependence' of particular development 'trajectories' leading to possibilities of 'lock in' for firms or regional economies (see, for example, Grabher, 1993; Arthur, 1994).

Perhaps the most lasting of value among the rich variety of Hirschman's concepts was 'linkage' (Hirschman, 1958). This, along with Dahmén's (1988) theory of 'development blocks' and Johanson and Mattsson's (1989) 'network' approach gave rise to the presently influential concept – in both policy and academic debates – of industrial clusters (Porter, 1990; 1998). In common with Perroux, Hirschman placed faith in the capacity of input–output modelling to identify key backward or 'upstream' and forward or 'downstream' linkages with the propulsive growth industry. These external economies would, he argued, be slowly induced by increasing demand from the complementary customer or user industry. Hirschman owed a debt to Veblen, a pioneer of institutional economics in identifying the effects of complementarity. Veblen, in a paraphrase that would have suited Schumpeter, stated that 'invention is the mother of necessity rather than vice-versa'. Hirschman's concepts of complementarity effects, induced demand and linkage gave considerably less hostages to fortune for regional development than the Cartesian determinism of Perroux's growth pole concept, or the rather mechanical assumptions of Keynes' accelerator, relating investment to previous increases in output.

Two other points about Hirschman's linkage analysis are worth recalling. The first is that in contrasting the relative importance of upstream and downstream linkages, he opted for the former because they most clearly reveal the pressure of demand, which is the essential dynamic of the growth process. In Porterian terms, upstream linkages would be higher up the *value chain* and thus more knowledge-intensive, stemming from applied research. These, in turn are key elements of competitive advantage because of their capability-inducing effect upon innovation, a key driver of productivity growth, hence competitiveness. A second point is that Hirschman also stressed the importance of not making abstraction the centrepiece of analysis. Rather, the particular conditions prevailing in a linkage system had to be examined in their regional or national context. This was not least because input–output tables, useful for descriptive purposes, are less so for analysing causation. This is due to key aspects of the linkage process not being available to the input–output technique. No information on innovation capacity, employment growth or institutional factors can satisfactorily be mobilized. Hence, the need for qualitative analysis, to enable sense to be made of the quantitative results of linkage studies.

Having noted the influence of Hirschman on Porter's thinking about industrial clusters, it is briefly worth mentioning the work of his relatively less well-known contemporary Erik Dahmén. His concept of 'development blocks' stressed linkage from development in one industry sector influencing development in another. In Schumpeterian terms, these show up particularly strongly when a radical innovation induces imitative and adaptive learning by the following entrepreneurs. Tracing these networks of linkage relations reveals the block of industrial sectors and firms that develop following application of the lead innovation and subsequent incremental innovations and adaptations. Andersen (1992) summarizes the evolutionary nature of the concept in four steps:

1   The development block occurs where industries have the potential for adaptability (i.e. no lock-in) and a tradition of entrepreneurship.
2   The core innovation causes a systemic shock increasing the disequilibrium of the economy in question. New market niches appear which may be filled by secondary innovations.
3   The size and shape of the development block can be defined after the shock has died down but is also preconceived by Schumpeterian entrepreneurs as the basis for their investment calculus.

4   An economy possessing development blocks that are both well established and industrially 'immature' is strengthened in its developmental potential.

Clearly, Dahmén's concept is also rooted in a disequilibrium perspective, Schumpeterian in inspiration and evolutionary in outlook. Examples of development blocks include, historically, the British textile complex, and contemporarily, British pharmaceuticals and financial services, German capital and intermediary capital goods such as industrial machinery for both heavy and light engineering markets and, on a smaller scale, Danish agro-industrial machinery. Newer blocks have emerged in US ICT and aerospace, for example.

There is, finally, a conceptual link from this type of analysis of economic evolution to 'product life-cycle' theory associated most closely with Vernon (1966). In straightforward terms this links innovation and regional economic development by postulating that a product starts its 'life' in the region where R&D is performed. It then migrates first to domestic production locations during the period when its commercialization means rapid growth in demand and the need for incremental innovations or adaptations. Thereafter, in its mature phase, when domestic demand has peaked, production shifts offshore, particularly if the producer is a multinational firm already, though a firm can become multinational via this phase of the process.

Two critical points can be made regarding the process envisaged in Vernon's hypothesis. One links back to the question raised by Hirschman and Porter as to the preferability of upstream over downstream linkage for development induced by Schumpeterian innovation. Andersen (1992) speaks here of a 'transformation problem' occasioned by user–producer interaction to debug the initial innovation. When this problem is solved, a second 'transformation problem' occurs between producer and suppliers. Out of both of these problems, the combination of pioneering and incremental innovation will be responsible for creating more niches where demand grows and incentives to profit are high. Thus the upstream, higher value-added stages of the innovation–commercialization process are the most attractive kinds of (upstream) linkages to have as a motor for a regional development block.

Second, however, the Vernon (1966) formulation is a linear diffusion model that simply assumes a core–periphery equation between innovation and exhaustion in the life of the product. Diagrammatically the 'line' is actually a logistical or 's'-shaped curve. What were earlier seen as 'transformation problems' are conceived by Robertson and Langlois (1995) as points of uncertainty. In theory the line declines towards the mature end of the logistical curve but it can rise at the mature phase when an innovation has a systemic impact on a mature product. This can cause a disruption to the 's'-curve, referred to as either mature uncertainty or mature growth. Examples would include the impact of microelectronics on mature products such as television sets, automotives or machine tools, or, in process terms, the introduction of 'lean production' (Womack *et al.*, 1990). Robertson and Langlois (1995) conclude that as long as these user sectors remain buoyant they will appropriate significant benefits from their upstream interaction with the innovators. Where they enter decay, however, both user and innovator suffer deficits in ways Florida and Kenney (1990) note for the case of the US consumer electronics industry.

Finally, it should not be overlooked that the kind of social interaction between users and producers or producers and suppliers around these points of uncertainty or transformation problems will shift according to the degree of uncertainty and maturity of the product. Where mature uncertainty is the problem, the key is to coordinate integration of the innovation that itself may be relatively well understood into the product quickly. But top-down *coordination* is inappropriate where the supplier of the innovation is in the early growth phase of production and uncertainty is high. There a *network* form of interaction is preferable because it maximizes the number of possible interactions between sources of potentially useful information. This is a condition that is most likely to be met in circumstances where there is a geographically proximate location of the range of actors involved in the interaction process supplying positive externalities into the decision-making process. Where, by contrast, developers and users are separated or where companies have become 'hollowed out' to the extent that they no longer have the capability to interact on specific innovations, economic efficiencies are lost. The latter issue of 'absorptive capacity' (Cohen and Levinthal, 1989) is one to which we shall return later.

To recap the argument that has developed thus far, we can say that classical political economy contained important insights into the processes of economic development, notably productivity gains from the division of labour, comparative advantage, trade specialization and agglomeration with external economies, many of which have great relevance for the analysis of regional and small-country economic development today. Then early regional economists added a further round of improvements by analysing the sources of localized economic growth whether in industrial districts or Weberian resource versus market trade-offs. But with Lösch the shift towards a neoclassical divorce from reality gathered pace as the pursuit of a fully mathematized general equilibrium theory began. However, a number of disequilibrium theorists, notably Schumpeter, offered an alternative analysis to the orthodoxy and one to which the work of Perroux, Myrdal, Hirschman and Dahmén can usefully be related. Much of this work returns us circuitously to the re-valuation of the regional dimensions of agglomeration, externality and specialization which are so crucial to contemporary innovation, hence growth and economic development.

The unification of equilibrium and disequilibrium approaches?

The disequilibrium tradition of Schumpeter and others produced important objections to equilibrium theory. This was mainly because of the metaphor or analogy which underpinned it from the outset in the work of Marshall, Jevons and Walras, the instigators of the 'marginalist revolution'. This laid the foundations for neoclassical orthodoxy. Drawing on physics they sought to equate economic functions with those of classical mechanics. From its outset, therefore, neoclassical economics has been based on a fundamental misconception that prevents it from explaining endogenously generated change. As Ulrich Witt puts it: 'There is no doubt, indeed, that neoclassical equilibrium theory has, *until recently*, had serious problems in explaining even what the theory itself had identified as the most prominent driving force of economic development: technical progress' (1991, p. 83; emphasis added). The reference to 'until recently' is important because neoclassical economists claim to have solved this and many related problems in their approach as a result of developing 'new growth theory'. And, with respect to the 'solution' of the problem of endogenous technological change, it is argued, it is the work of Romer (1990) among others that claims credit. We shall return to this claim below to assess its validity.

It is first necessary to explore further precisely why the analogy at the heart of neoclassical economic theory is so misleading, a feature which theorists more persuaded by an evolutionary perspective hold to pervade also the work of neoclassical 'new growth theorists'. By specifying foundational weaknesses we can proceed to an exploration of alternative perspectives that seek to eschew reductionist conceptions of human nature. Witt continues his observations on foundational weakness as follows:

> In mechanics, disequilibrium is caused when some outside force creates free energy within the system under consideration. The first element of the neoclassical hard-core synthesis – the constrained utility maximization hypothesis – carries this basic concept over to economics. The economic analogue to free energy is psychic energy, drive or motivation to act on the part of the individual.
>
> (1991, p. 86)

From this is derived the measure of utility as the expression of individual choice or preference between the present condition in regard to wants or desires and those that may be exogenously presented. When that calculation has been acted upon, the externally derived 'energy' disappears.

It is well known that a further, crucial assumption of 'the constrained utility maximization hypothesis' is that the actor in question has perfect knowledge of the range of possible preferred alternatives and a capability to anticipate by thinking through their broad consequences. No room is left for the possibility that, by introspection, the actor might come up with a set of preferences that are not exogenously given. Thus, when, simultaneously, all individual utilities have been maximized, there occurs an equilibrium, a perfect arrangement of wants – satisfaction from which no deviation need rationally be sought. Even taking into account the relaxation of such questionable notions as perfect information by introduction of modifiers such as 'satisficing' or 'second-best' solutions, the underlying misconception remains. That is, for the schema to work, humans must be conceived as cognate to Pavlov's dogs; human action is stimulated only by exogenous causes, hence no free will – an ironic implication for a theory formed by the idea of markets as precisely the best mechanism for realizing freedom of choice.

Newer economic thinking from inside and outside the neoclassical paradigm has been perplexed or even further dismissive of such paradoxes. Those inside the paradigm have sought to develop a better theorization of neoclassical economics itself while among those outside it, there has been the effort to develop a superior alternative called evolutionary economics. But as Witt (1991) notes, the latter have been better at critiquing the former, and especially their forerunners, than developing a consensus on core concepts of the alternative paradigm themselves. We shall see in the last section of this chapter what progress there has been in the latter venture, first, it is instructive to look at some of the progress that has been made by the 'new neoclassicals', notably Paul Romer and, particularly, Paul Krugman.

New growth theory and the new neoclassicals

Turning to neoclassical economic growth theory, we find two key assumptions, both of which have proven problematic. The first is that the production function – the master concept linking economic output to labour and capital inputs – operates under conditions of diminishing returns. In other words, adding capital to labour or labour to capital will result in consecutively smaller increments in output. Second, neoclassical growth theory assumes constant returns to scale or the notion that increases in inputs result in an equivalent increment in output. An obvious problem with both assumptions is that they cannot explain a common feature of modern economies, namely, increasing productivity per unit of labour. The exogenous impact of technological change was deployed in the neoclassical approach to deal with this lacuna, to the dissatisfaction of many of its adherents. This dissatisfaction gave rise to 'the Solow residual' (Solow, 1957) who distinguished between labour productivity, capital productivity and a residual – measured by technological change. Empirical research conducted to test this theory concluded that in some empirically measured economies as much as 80–90 per cent of growth is accounted for by improvements in technology.

Technology is part capital and part know-how. Neoclassical growth theory holds that investment in capital is the vital ingredient for stimulating growth in less developed economies. Their expected lower labour costs would mean a higher return on capital because, under the diminishing returns assumption the relationship of labour to capital costs is constant. So there should be a premium for investors from a higher, though ultimately diminishing, rate of return on capital in the less developed compared to the more developed economy. Given such investment, higher wages (labour costs) should eventually follow, thus producing equalization and inter-economy convergence in growth indicators, or equilibrium. The evidence for any of these propositions actually occurring in reality is hard to find. To be sure, capital flows to less developed economies, but far more (over 80 per cent of foreign direct investment, according to Ruigrok and van Tulder, 1995) flows between the developed countries. The Newly Industrialized Countries (NICs), the strongest empirical props for neoclassical growth theory, accounted for about 10 per cent with an annual growth in share of global industrial production of 0.1 per cent (1973–84) (Gordon, 1988). Equally, few commentators perceive a strong convergence but rather a marked divergence between income levels in less and more developed economies. This is one reason for the vast increase in migration from the former to the latter during the closing decades of the twentieth century.

Owing to these debilitating weaknesses, a key focus of new neoclassical economic literature, referred to as 'new growth theory', was upon *endogenous* growth. Not surprisingly, the large 'Solow residual' explained by technological change was of primary interest. Two areas of particular relevance are those involving, first, human capital and, second, technology. With respect to the first, Lucas (1988) identified the external economies that arise from human capital, i.e. the skills of workers, especially with respect to the capacity for 'learning by doing' (what might be called 'on the job training') as potentially important and worthy of modelling. The reference to externalities is of direct relevance to our earlier discussion of both equilibrium and disequilibrium approaches to economic development. This, as we shall see later, is where the presence of scarce labour pools attract entrepreneurship and inward investment. The Thames Valley in the UK attracts software firms like Microsoft and Oracle because there is a pool of scarce human capital expert not only in writing code but selling products. This, like the nearby motor sport cluster centred on Oxford which concentrates the world's leading racing automotive expertise and all the leading Formula 1 design teams, is an example of one of the spillovers that give clusters advantage as labour moves from firm to firm on short-term contract. Lucas points to the external effect of human capital on overall productivity because of its imperfect appropriability. That is, the effect of an increment of skill pervading its market is greater than its value to the 'owner' of that enhanced skill. Among other things, Lucas says, this helps explain the existence of cities, which are a particular form of often multi-layered, overlapping cluster. These are not low-cost locations for business yet they attract it nevertheless. This is justified because of the productivity increase arising from human capital externalities. Moreover, he says that urban land rent measures this effect in value terms. Further, there is a cumulative benefit to this process whereby the higher the level of learning embodied in the interactions occasioned by co-location, the faster will be the growth rate of business in that location.

Clearly, the Lucas approach yields a neoclassical equilibrium model of cumulative causation, after Myrdal and Hirschman. Hence it redescribes the polarization process endemic to capitalism, the new twist being that the key factor in that process is the cumulative level of capacity for learning by doing. This is further bolstered by the thesis that such learning engenders externality effects that are themselves crucial locational attractors in the economic development process. The main problems arising from the analysis Lucas makes is that it implies the higher the level of learning, the faster the growth of the economy, thus preserving the neoclassical mythology of, in this case, perfect transmission of learning in externality form. But, as we have seen, this assumption is rather heroic in contexts where spillover effects are less 'in the air' than controlled, perhaps only temporarily, by an in-crowd.

Romer's (1990) critique of Lucas is that he treats knowledge as an unanticipated public-goods externality of the normal production function and thus neglects individual autonomy in knowledge generation. In other words, technological change is treated as something purchased off the shelf. His 'new neoclassical' approach involves a broader conception of the centrality of technological change as itself an improved production function, something indigenous to the firms that use knowledge to raise efficiency, such improvements intended by market-motivated individuals, the effect of whose decisions is to produce increasing returns to scale. Increasing returns to scale are the cumulative rather than merely marginal returns that accrue to first-mover firms in a knowledge-based industry or those that otherwise gain a monopoly position in a given market. Some Knowledge Economy firms with an arm-lock on key Internet technologies, like Cisco Systems, virtual monopolists in Internet 'routers' which control Internet communication channels, have this characteristic, as did telecommunications monopolies before privatization and the introduction of competition for telecom services market-share. Increasing returns arise from the widespread and repetitive gains that accrue from the continuously improved codified knowledge responsible for the technological change in question. In Cisco's case this is achieved by acquisition of specialist knowledge-intensive smaller businesses in appropriate technologies. In other cases it arises from equity investment by corporate venturing arms of such (quasi)-monopolists in equivalent small businesses, as in the case of Intel Technology, which is assumed to possess this increasing returns characteristic. This, in turn, requires the replacement of the neoclassical cornerstone assumption of perfect by one of imperfect competition, since increasing returns to scale clearly imply monopolistic rather than fair competition. While the work of Romer, like that of Lucas, marks a welcome move towards a recognition of reality's tendency to economic divergence rather than convergence in economic growth, its neoclassical origins still give problems that neo-Schumpeterians, especially, find overwhelming:

> [M]ost of these models suffer from some of the same unrealistic assumptions and the same measurement problems as the 'old' growth models. In particular, they take little or no account of organizational innovations and of the interplay between institutional change, technical change and investment.
>
> (Freeman, 1994, p. 85)

To which could be added that there remains a strong tendency to overestimate the 'absorptive capacity' of firms where both human capital and technological change are concerned (Cohen and Levinthal, 1989).

Krugman's exploration of the spatial question

In short books such as Krugman (1991; 1995) the author has moved a little closer to recognizing some elements of Freeman's critique of new growth theory. In Martin and Sunley's (1996) assessment of his work they identify trade, externalities, industrial location, industrial strategy, globalization, history and path dependence. Above all, Krugman has sought to understand what he sees as the fundamental importance of geography for economic theory. In particular, he has, along with the other new neoclassicals, aimed at explaining regional disequilibrium by reference to Marshallian externalities, regional science and, like Lucas, cumulative causation. For this he is, rightly, congratulated by some economic geographers but inevitably, dismissed by others. While Lucas and Romer made their contribution to 'new growth theory' via human capital and technology analyses respectively, Krugman's foray started as 'new trade theory'. Moving beyond Ricardo and Heckscher–Ohlin, he focused on a key lacuna in traditional and neoclassical trade theory, which was its inability to explain intra-industry trade. That is, in contemporary economies, the fastest growing trading is between economies exchanging similar rather than different goods and services. Examples would be luxury cars from Europe being exchanged for off-road vehicles from the USA in trade between the two economies. Like Romer, Krugman buys into the assumption that increasing returns to scale, hence imperfect competition, underlie this. A key implication of this is enhanced specialization of economies in their strongest industry because of cumulative causation or 'lock-in', and path dependence occasioned by increasing returns. This means it is difficult to move from a winning, but also in time, perhaps, a losing trajectory arising through increasing returns from a monopoly market position that created an industry monoculture. Many Old Economy industries face such 'lock-ins', as do regions that sustained such monocultures. Sustaining this initial advantage are technological supremacy and strategic trade policies, increasingly capable of being fashioned not only, as in the past, by states but also supra-states such as the EU, and regions possessing the appropriate governance capability. Competition rules by the World Trade Organization, the EU and economically liberal governments have lessened 'neo-mercantilism' of the trade-protection kind, but increasing returns raise serious questions about policies based on specialization rather than diversification. Support for R&D and constant innovation is one sphere left free of competition rules for now and the foreseeable future because of the high risk of investment and the low appropriability of successful results of commercialized research and innovation. Retro-engineering is always possible.

At the regional and local levels Krugman explains the formation of regional agglomerations as the interaction of external economies of scale under conditions of imperfect competition, including Weber-like transportation returns to scale. Where the latter are low, concentration follows, where they are high, dispersion is the result. Innovator locations will benefit cumulatively from that good fortune, in the process, pulling investment away from less-favoured regions. Hence, like Lucas, Romer and the earlier disequilibrium theorists,

Krugman appears to be a theorist of cumulative inequality between regions and countries. On the other hand, his emphasis on specialization and intra-industry trade leaves a space for growth to occur through trade as specific, regional, industry clusters become significant through their accumulation of complementary forward and backward linkages and competitive advantage. Interestingly, and unlike other new neoclassicals, Krugman recognizes that substantial change, even reversal, in the development trajectory of a region is entirely possible though unpredictable.

Consistently from his 1991 through to his 1995 lectures he argued that the reason both equilibrium and disequilibrium regional science failed to retain a position at the high table of economic theory was a failure to comprehend and deal with the problem of market structure (i.e. monopoly). That is, regional science and its predecessors

> must implicitly involve some strong beliefs about market structure ... but the story is entirely implicit, and had to be so, since monopolistic competition was not something people knew how to model in the heyday of the market potential approach.
>
> (Krugman, 1995, p. 46)

Now, as we have seen, neoclassicists have developed models to do this. He goes on to argue that in the world of 'new growth theory' with imperfect competition, increasing returns to scale and technological externalities or spillovers, agglomeration, involving intermediary firms supplying linkages to customer firms, produces precisely the kind of market structures which give competitive advantage, particularly to regions and their key clusters. But the uneven distribution of advantage that accrues from this basically unequal process is also the cause of regional disadvantage and its associated disparities; intra-industry trade is, for Krugman, the main means of salvation, but on the question of time-scale he is silent.

Krugman's work on economic geography understandably develops as more is assimilated but he cannot be said to have left the neoclassical camp. Rather, as we will see, he assimilates non-neoclassical theory into a formally coherent neoclassical modelling framework. This is suggested by Martin and Sunley (1996; see also Martin, 1999) who distinguish two different models that Krugman advances in explanation of regional divergence. The first is based on capital flows to low labour cost regions which are outweighed by market attraction back towards the donor region, advantaged by its external economies including advantageous transport costs. The second is that because of increasing returns and their consequential *specialization* effects, not only will regions become locked in to divergent paths, they will also be more prone to economic booms or slumps. This will produce a regional economic landscape of increasing-returns advantages from particular specialization innovations within intra-industry trade, quickly followed by decreasing returns disadvantages, as the erstwhile innovator region is transcended by the most recent one. The sense of a clock ticking as the scanning of the minute hand coincides hourly with a different number (region) is only dissipated by the realization that this is a random process.

But this is precisely where neoclassicism lets Krugman down. Essentially what he does is stress imperfect competition, increasing returns and externalities, technological or pecuniary but treat them as if they were perfect competition, constant returns and internal economies of scale. In support of this, Martin and Sunley (1996) quote Krugman's explanation of how he does it as follows: 'What look like highly irrational outcomes in the marketplace are caused by the interaction between imperfectly competitive markets *and slightly less than perfectly rational individuals*' (Krugman, 1994, p. 213; emphasis added). It is this 'slightly less than perfectly rational individual' point that suggests the strongly neoclassical foundations of Krugman's project.

This becomes even clearer from perusal of his volume of three essays based on his Ohlin lectures given at Stockholm (Krugman, 1995). There, entertainingly as he hopes, he does three things. First, he asks why development theory, so powerful in the 1950s disappeared from the high table of economic theory thereafter. The reason he advances is its inability to model increasing returns and externalities and thus market structure. Second, he asks why regional science has never occupied a seat at high table. Reviewing most of the work presented as equilibrium and disequilibrium theory in the first two sections of this chapter, he concludes that they are basically identical in being good at explaining 'why economic activity spreads out, not why it becomes concentrated ... What economic geography ... needs ... is a synthesis that brings back the other half of the story' (Krugman, 1995, pp. 54–5). But, equally, they lack the capacity to deal with market structure or monopolistic competition.

It has to be admitted he has a point here and he goes on to map out a model for doing what he says geographers have failed to do. Their failure to understand market structure, is addressed by showing how his model generates a cumulative process of multiple agglomerations with Löschian central place outcomes. The metaphor the model adopts is indeed that of a clock with different iterations causing externalities to favour a two-agglomeration regional system. Here, forward and backward manufacturing linkages are influencing the model dynamics. He adds two important notions, one in explaining centrifugal developmental tendencies by the (rational) desire of producers to escape competition (i.e. create a new 'more perfect' market), the other he calls the 'agglomeration shadow' which is cast by the more dynamic agglomerations, thus limiting the growth of competitors. In other words, clusters are monopolies in the market for economic space. The third thing he does is, once again, strongly defend mathematical modelling as the only way to be taken seriously in economics.

This is still an (imperfectly) rational utility-maximizing model in which trade-offs are made between agglomeration and non-agglomeration or space versus transportation cost indices as firms seek the most rational outcome in an imperfect market economy. By unifying equilibrium and disequilibrium approaches in the way he does, Krugman develops what might be called a 'respiratory' model of regional economic growth in which competition is expressed in processes of cluster concentration and deconcentration. As Krugman puts it:

> Any interesting model of economic geography must exhibit a tension between two kinds of forces: 'centripetal' forces that tend to pull economic activity into agglomerations, and 'centrifugal' forces that tend to break up such agglomerations or limit their size.
>
> (1995, p. 90)

But this sits uneasily with his thesis that, in the earlier formulation, Myrdalian backwash effects will predominate over spread effects and, in the later one, that intra-industry trade specialization will produce unpredictable booms and slumps for specific agglomerations.

The absence of any capability in modelling to handle the countervailing effects of regulation, institutional evaluation, organizational action, history, interactive learning and the qualitative not merely the quantitative dimensions of externalities means that Krugman is condemned to rediscover, albeit with better mathematical proofs, problems rather than identify solutions or new problems of economic geography. It would help were he to expand significantly, along lines pioneered by the likes of David (1985) and Arthur (1994) his understanding of 'lock-in' since that is precisely one of the features which produces the persistence rather than the demise of cities and even of regions that may have become competitively disadvantaged. In this respect, his deductions are similar to those of Harrison (1995) and others who saw no logic in the thesis that clusters can survive and be regenerated over lengthy periods. 'Lock-in' is part of the positive as well as the negative story of clusters because it is the capability to innovate within exclusive networks that provides the key that locks in learning capacity. Only when that learning has been superseded and new knowledge not absorbed does 'lock-in' become a problem.

Evolutionary economics: a new heterodoxy

The inadequacy of the new neoclassicals' treatment of time, the interaction of human agency with institutional constraint and opportunity, the very creative processes that stimulate initiative-taking, the nature and complexity of learning not only new but how to re-use old knowledge and the subtleties of agglomeration economies and diseconomies are not reproduced, but rather made central objects of study in evolutionary economics. That these are crucial to the understanding of economic development processes is testified to not least by the energy with which the likes of Lucas, Romer and Krugman have sought to integrate them into neoclassical economics though, as we have seen, with only a moderate contribution to furthering our understanding. Interestingly, in passing, Dymski (1996) argues that Krugman has consistently sought to distance himself from 'the Mainstream' of neoclassical general equilibrium economics but largely fails in that enterprise, despite inventing a 'Mainstream 2' which he firmly distinguishes from heterodox approaches such as political economy or evolutionary economics.

As Witt (1991) sees it, there are four distinct schools of economic theory that converge into what is now called, after Nelson and Winter's (1982) seminal work, evolutionary economics. They are: (1) the Schumpeterian tradition; (2) institutional economics; (3) Marxist political economy; and (4) the Austrian School of Hayek and others. Key elements of the Schumpeterian and neo-Schumpeterian approach have been noted already so at this point a summary only is presented. The first, most important point to make here is that Schumpeter recognized very early the centrality to economic growth and change of the impact of innovation in products and the organization of their production. Second, as we have seen, the 'heroic' role of innovators, initially as entrepreneurs, later as R&D engineers was perceived as being crucial to his theorization of economic development. Their skill is to unite the findings of science with opportunities for profitable investment, signs of which attract imitators engaging in incremental innovations in the wake of the radical efforts of the initiators. Third, this process unleashes a wave of 'creative destruction' of investments based on earlier rounds of innovation. Thereafter the innovation would be expected to diffuse and its products move through a life-cycle until replaced, in turn, by the new wave. Innovations might occur unpredictably in different industrial sectors but the radical ones would be likely to have a pervasive effect, initiating an upturn in the business cycle.

Neo-Schumpeterians have developed many of these insights and sought to investigate points of criticism, such as that cited by Freeman (1994) to the effect that Schumpeter's approach lacked a theory of innovation as distinct from a theory of entrepreneurship. Neo-Schumpeterian economists and others have made progress in five areas. First, they have complemented Schumpeter's findings on originators of innovations with radical impact by showing the importance of *incremental* change in the innovative performance of firms. Second, it has been found that exogenously derived *scientific* findings, especially when embodied in new recruits from basic research to firms, are an essential innovation growth factor in successful firms. Third, the importance of social interaction or *networking* relationships between innovation actors in a multiplicity of organizational settings has been identified as being associated with successful innovation. Fourth, the role of sophisticated *users* of innovation in the process of product or process development has been shown to be increasingly a feature of successful innovation. Fifth, the existence of *systemic* and integrated relationships between the research, engineering (product or organizational), marketing and related firm functions is most often associated with firms demonstrating capability successfully to innovate. Above all, these combine to highlight the central value of a *learning* disposition to firms whose innovation processes have been analysed from within a broadly neo-Schumpeterian perspective.

It is plain to see that most of these key findings of recent theoretical and empirical research into innovation are presenting a portrait of this crucial part of the economic development process as distinctly evolutionary in nature. The emphases upon *incremental* change as the element of 'normal science' *à la* Kuhn (1962), framed by the revolutionary punctuation marks of radical innovation and *learning* whether from science, users or other business professionals point to the iterative and approximating processes by means of which economic change may be perceived to occur. In complementary fashion, the emphases upon *networking* and *systems integration* point to the fundamentally social processes of information transmission and diffusion through interactions between agents of consequence to the innovation element of economic development. Both of these are to be contrasted to the hierarchical, command and control strategies of big science associated with linear thinking of a more planning-conscious era when making 'breakthroughs' was seen as the key innovation goal. This is the kind of work Schumpeter invited his successors to pursue in light of his observation that: 'The essential point to grasp is that in dealing with capitalism we are dealing with an evolutionary process' ([1942] 1975, p. 82).

If the neo-Schumpeterian school has played the central role in promoting evolutionary economic theory and research to a leading position among the challengers to neoclassical orthodoxy, it is closely attended by its sister school of institutional economics. This field received a significant boost in 1995 when one of its key proponents, Douglass North was awarded the Nobel Prize for economics. His writing on the subject is sufficiently clear to serve as a valuable introduction to many basic elements of this approach. In his contribution to the collection edited by Mäki, Gustafsson and Knudsen entitled *Rationality, Institutions and Economic Methodology*, North (1993) lays out important elements of the institutional perspective with their emphasis being particularly towards issues of *trust*, *institutions* and *cooperation* in economic performance. North's definition of *institutions* is classical, equating them with rules and norms of behaviour and the way they are enforced; in other words, 'structured human action'. The structures or means of enforcement include *conventions* which he sees as solutions to simple or everyday problems of coordination in the way that 'keeping your word' is an indicator of honesty or integrity, enabling transactions to be costless and reliably made. Where more complex forms of exchange are involved, *rules of the game* define the extent of transaction choice open to economic actors (e.g. rules disallowing 'insider dealing' in stock market transactions).

Institutions thus have the following economic effects: first, they shape the incentive structure, which, in turn, determines economic growth and hence the way economies evolve historically. Second, they provide the framework of rules and constraints that define the opportunities for enterprise in an economy. Third, they set the scene for *organizations* such as trade unions, firms and political bodies to seek to survive by developing their learning capacities (skills, knowledge, etc.), which, in turn, exert their evolutionary effect on the shape of the economy. Finally, the institutional processes described, evolving incrementally because of innumerable interactions by political and business actors, exist to manage the expensive practice of structuring exchange by means of transaction costs. Even in 1970 transaction costs accounted for nearly 50 per cent of US GNP and thirty years later, given the growth of litigation and the extent of externalization to suppliers, it can be expected to be higher. By creating less uncertainty in the economic environment, institutions may assist in the reduction of these overheads.

North makes interesting points concerning the role of cooperation in economic exchange when he draws attention to the following finding from game theory, a method of simulating human negotiating behaviour: 'Wealth maximizing individuals will usually find it worthwhile cooperating with other players when the play is repeated, when they possess complete information about the other players' past performance, and when there are small numbers of players' (1993, p. 244). Because these conditions are rare in modern economies and the more complex forms of exchange typically found there invoke high transaction costs, competition is the norm. However, where even more complex exchange is susceptible to a successful resolution of interests, institutions that enhance the benefits of cooperation or the costs of defection may be created. The institution of 'collective bargaining' between employers' and employees' organizations is an example, as, before they were outlawed in the interests of market competition, were price-fixing cartels. Modern-day 'strategic alliances' of a 'pre-competitive' nature have much of this *institutional* character about them, though where they are fairly informal, information-trading understandings, they may take on more of the character of *conventions* among firms.

Finally, why and how does institutional change occur? First, the key functions of institutions are to secure stability and continuity thus enabling complex exchange to be facilitated across space and time, a process in which 'the status quo gets favoured' (North, 1993, p. 252). Second, informal constraints such as routines, customs, traditions and culture add further stability allowing an 'autopilot' character to economic behaviour. But change occurs, third, to institutions because of shocks such as significant shifts in consumer preferences or prices, though incremental change is much more normal, cumulatively creating a long-run path of change.

Thus, as North puts it in a much-quoted sentiment: 'Path-dependence means that history matters' (ibid., p. 256). Hence economic *growth* occurs where the institutional structure rewards productive activity, while economic *decline* persists where there is a lack of feedback to create organizations that give incentives for productive investment. And in a passage reminiscent of new growth theory, North concludes that: 'Indeed it is the increasing returns characteristic of an institutional matrix that produces path dependence' (ibid., p. 256). In other words, for North, economies are mainly slowly-evolving entities overwhelmingly characterized by incremental change. Even when discontinuities occur, formal rules may change but informal constraints less so and there is eventually an evolutionary accommodation between the two.

Clearly, North's version of economic evolution is more conservative than, though bearing some partial similarity with, that of the neo-Schumpeterians. He stresses continuity far more than discontinuity, indeed the theory of institutional change he evokes is rather limited to neoclassical variables such as consumer preference or price, though what lies behind either shift is left somewhat unclear. Similarly, even when he accepts, as he does, that revolutionary change can occur in economies – not least where political revolutions take place – even these are seen as being trapped by the 'short half-life of the ideological consultant' (ibid.) such that the end result is a blend of pre- and post-revolutionary institutional constraints. There seems accordingly to be little room for the impact of other than incremental innovation in North's scheme and this must be considered a limitation even, or perhaps especially, in the long run, which is his professed speciality.

Most reviews of institutional economic thinking begin with an appraisal of the pioneering work of Thorstein Veblen as in Veblen ([1899] 1976). As Hodgson (1996) puts it: 'Veblen rejected the continuously calculating, marginally adjusting agent of neoclassical theory to emphasise inertia and habit instead ... Institutions thereby help to sustain habits of action and thought' (1996, p. 408). But Veblen was probably less of a determinist than North became, since he recognized the relevance to economic change of creativity and innovation while rejecting the notion of institutions as primarily constraining structures. Nor was he interested in the notion that individualism was determined by a person's social or class position. Equally neoclassical 'economic man' assumptions received short shrift. Rather, Veblen thought individuals and the firms which they compose had strategic capacity and alternative routes to achieving their objectives. These give rise to culture which differs from firm to firm. The institutionalized firm culture was formed cumulatively by the combination of individual adaptation of means to ends in a changing environment. This notion was the origin of Myrdal's (1957) elaboration of the theory of cumulative causation. This in turn, as Hodgson (1996) points out, and we shall explore further in Chapter 4, gives rise to contemporary theorization of processes of 'path-dependence', firm or regional 'trajectories' and 'lock-in' (Arthur, 1994). This is at odds with the traditional neoclassical notion of the pre-determined tendency towards equilibrium in economic development. The combination of the idea of individual autonomy and cumulative causation means there are different regional trajectories in economic development.

Veblen's successors constituted an important school of economic thought as they explored institutional insights. Commons (1934), for example, developed the idea of diversity in the pathways to economic development arguing, among other things, that the history and institutional structure of the US economy meant that it developed economic development organizations such as its trade unions, differently from those found in European economies. Commons also followed Veblen in stressing the opportunity for creativity or liberation along with the constraining effect of institutional structure. Thus the notion of distinctive solutions to the problem of cumulative causation arising from institutional and cultural distinctiveness of a regional or local kind continues in this tradition. This may include socio-cultural conventions such as associative civicism as will be explored in Chapter 3.

So the institutionalist approach to economic theory resonates with a disequilibrium perspective. There is recognition of the centrality of disruptive change to the institutionalist perspective, which is shared with that of neo-Schumpeterian analysis but just as some of the latter has its conservative emphases, so in the work of North and Williamson, does the institutional school. However, drawing attention to the structure–agency dialectic and refusing to foreclose economic development as an inevitably self-balancing or self-destroying process is useful. In recognizing economic decision-making as involving monitoring, iteration, approximation and learning by actors solving problems creatively, innovatively and unexpectedly, the institutionalist school offers much to understanding of imbalance in the economic development process, one form of which evolves into clustering practices by firms.

Two further strands of economic thinking demanding attention in this attempt at exposition of the features of the evolutionary economics approach are the schools of Marx and Hayek. Ideologically and epistemologically polar opposites with their championing of collective socialism and market individualism respectively, it is perhaps only their reductionism that unifies them, apart from their distinctively different evolutionism. Schumpeter noted that capitalism is an evolutionary process and continued:

> It may seem strange that anyone can fail to see so obvious a fact which moreover was long ago emphasised by Karl Marx. Yet that fragmentary analysis which yields the bulk of our propositions about the functioning of modern capitalism persistently neglects it.
>
> ([1942] 1975, p. 82)

The point here, which Schumpeter learnt from Marx, is that capitalism can never be stationary. The key force creating this instability is itself capitalist enterprise, its products, organizational methods, markets and forms of transportation.

Marxian analysis has no place for the theoretical or, for that matter, actual individual; it is an analysis based fundamentally upon the social category of class. That relationship in all hitherto existing class societies is one of exploitation. Ancient classes exploited the slave class; feudal classes exploited the peasantry; the capitalist class exploits the working class. Therein lies their source of wealth and power and, for the exploited class, their relative impoverishment. Marx had little to say about cultural or spatial variation in this, though some neo-Marxist analyses devote a great deal of attention to the ways in which national or regional economic systems are exploited, because of their distinctive (peasantry or working class-dominated) nature by other nations (Amin, 1976) or other regions (Lipietz, 1987). In this universalistic analysis, development and change occur when new technology and industrial organization come into contact with less developed modes of production. Critics such as Hodgson

(1996) point out that this very generality is a weakness because it assumes away the possibility of different evolutionary pathways to development of the kind insisted upon by the institutionalists, whether radical or conservative.

A second failure of the Marxian perspective is its assumption that there are non-capitalist institutions in capitalist economies But the 'gift' or rationally questionable economic relations surrounding the family, sacrifices made in support of culturally important icons whether the survival of an art form or of a language, or the often 'uneconomic', in rational terms, existence of people's support for a religion by virtue of the tithe or donation, make this a questionable assumption to say the least. The state itself is everywhere in deficit, yet impossible to bankrupt.

But, finally, in terms of economic development, the Marxian view is fundamentally flawed in its deterministic view of economic evolution. The stages of development are mapped out, as ancient modes of production give way to the feudal, itself destroyed by revolution to produce the bourgeois or capitalist form of economy and thereafter by revolution, socialism. Empirically, history has usually departed from this schema and where, telescopically, it occurred in practice in Russia, it failed precisely because of pre-existing uneven development which, when dragooned into the template of socialist centralism, destroyed the creative impulses which made life worth living. This is entirely a consequence of the absence of the concept of human individuality in the Marxian conception of the economy and society.

Not quite the same can be said of Hayek and the Austrian School of economists. He and they offer a concept of the individual at the heart of their project, though it is an attenuated and ahistorical one that they propose. Hodgson (1996) points to two versions of this caricature in Hayek: first, a kind of bond-trading pit where actors collide in pursuit of profits but which is institutionally unregulated. Second, Hayek presented a view of markets as an evolved social order among individuals, again in the absence of any institutional underpinning.

Whereas for Marx, capitalism is the abstract, universal principle, for Hayek, it is the market that holds theory, if not practice, together. Like Marx, again, Hayek sees market economies as self-sufficient and oblivious to non-market institutions and organizations. Thus the state is seen as an impurity that must be extruded for markets to function properly. Whether this is to be the fate of family, culture and church remains unclear. This, of course, is an absurd position, ignoring the economic history of the state's (and the family's, let alone the church's) role in securing market hegemony, as ably shown by Polanyi (1944) for Britain even, or particularly, at the height of so-called *laissez-faire*.

So why does the Austrian School and Hayek, in particular, have any relevance for an evolutionary perspective on economic development? In Hodgson's terms this is because:

> The Austrian school of economics has always given more emphasis to the explanation of the nature and evolution of socio-economic institutions than

most economists in the neoclassical camp. One of the classic cases here is Carl Menger's celebrated theory of the 'organic' and spontaneous evolution of money out of a barter economy.

(1996, p. 396)

Money 'emerges' as a preferred means of exchange to others by a 'convention' (cf. North, 1993) that it is the more reliable, trustworthy, convenient and portable medium of those available. It thus exerts a pandemic effect, which results in it being institutionalized, meaning accepted by everybody as the common currency. While historically dubious, this analysis is, in some ways, abstractly persuasive. Money was usually imposed by conquering cultures. Hence, the Austrian School is non-deterministic in its analysis, history develops from the interplay of individual actions, though with indeterminate motivations. However, it is unclear, as it is not with North and other institutionalists, where the limits on human liberty lie, this absence being celebrated ideologically by Hayek in a view of humanity which appears to reduce it, at best, to a kind of 'engine of desire' as some postmodernist sociologists have expressed it (Kellner, 1988).

To understand economic development in terms of imbalance and disruptive, spatially focused change has involved reviewing and seeking to synthesize different approaches. However, from this a number of foundational elements emerge. There are seven of these described below and which contribute to an evolutionary perspective on economic development in general and cluster development in particular.

First, the kind of approach taken here is interested in *processes* of growth and change of diverse kinds, in differentiated economies with distinctive cultural frameworks and political systems.

Second, it gives purchase on processes responsible for *comparative advantage*, especially as based on non-natural resources such as those involving knowledge, information and learning. Conversely, it is susceptible to explaining relative economic *disadvantage* in relation to spatial developmental processes,

Third, this approach illuminates mechanisms involved in processes of *agglomeration*, including the external economies, linkages and networks that influence the localization strategies of firms. Thus issues relating to labour markets, specialized inputs and technological or organizational innovation, the form and content of agglomeration economies, clusters, development blocks, and so on are of central interest.

Fourth, *innovation* in itself is a key process concern and interest, along with individual and group dimensions of creativity, including Schumpeterian 'creative destruction', radical and incremental innovation and entrepreneurship at the individual, enterprise at the group levels.

Fifth, the preferred perspective recognizes *disequilibrium* as well as equilibrium, cumulative causation, the centrifugal and centripetal processes by which economic development gains spatial expression, and the mechanisms underlying these.

Sixth, the approach is *non-mechanical* in its metaphorical imagery but will also be sceptical of overbalancing into full-blown biological analogy. It pays attention to the fruits of important revisions in neoclassical economics such as endogenous growth, imperfect knowledge and increasing returns to scale and production, trade and locational specialization.

Seventh and finally, the evolutionary approach to understanding these processes encompasses *institutional* and organizational matters often absent or marginal to orthodox economics discourse including institutional interactions, learning, trust-building, network management social capital, cooperation, diverse economic trajectories, path-dependence and 'lock-in'.

In taking a process approach, this perspective is valuable for understanding clustering practices and policy interventions that assist them. To that extent, such conditions as these find an echo in Hodgson's review of 'the evolution of evolutionary economics' where he delineates its grounding as follows:

> For example: a rejection of 'typological essentialism' in favour of 'population thinking' [i.e. replacing notions such as 'rational economic man' with more diverse conceptions of human behaviour], the recognition of the openness of socio-economic systems, the creativity and indeterminacy of human agency, the potentiality for novelty, the possibility of emergent properties, and the resultant undermining of reductionist methodologies.
>
> (1995, p. 482)

Hodgson also warns against the temptations of borrowing too freely from biological analogies such as those which compare, say, the routine of the firm with the genetic composition of species, even if such conjuring may have pedagogical benefits in the short run (see, for example, Nelson and Winter, 1982).

Andersen (1995) presents an elaboration of what he calls typical assumptions and characteristics of evolutionary–economic explanations that are compatible with the conditions outlined earlier and the grounding features noted by Hodgson (1995). They are as follows:

1   The agents (individuals and organizations) can never be perfectly informed and they have (at best) to optimize locally rather than globally.
2   The decision-making of agents is normally bound to rules, norms and institutions.
3   Agents are to some extent able to imitate the rules of other agents, to learn for themselves and to create novelty.
4   The processes of imitation and innovation are characterized by significant degrees of cumulativeness and path dependency but they may be punctuated by disruptive change.
5   Interactions between agents are typically made in disequilibrium situations and the result is successes and failures of commodity variants and method variants as well as of agents.

6   The processes of change occurring in a context described by the above assumptions and characteristics are non-deterministic, open-ended and irreversible.

Taking this approach and combining a neo-Schumpeterian analysis of innovation and the business cycle with cooperative game theory after Axelrod (1984), Andersen is able to demonstrate that it is cooperative behaviour among individual firms that explains the capitalist 'boom and bust' cycle as a process characterized by the two master-characteristics of '*punctuated evolution*', i.e. the economic system reveals long periods characterized by equilibrium and short periods of disequilibrium, and '*jerky innovation*', i.e. the process whereby in successive iterations between producers of innovations and their users, on the one hand, and suppliers, on the other, a 'new combination' or innovation is eventually introduced to the market, which then releases the swarm of innovators seeking to capture early innovation rents. Andersen also sees this as a highly clustered process in which a great deal of tacit knowledge is being exchanged around any or all of Schumpeter's five kinds of innovation: product, process, input, region and organizational form.

Conclusion

Large strides forward have been made in theorization of the problems of growth, change and economic development. Some of the deadening hand of neoclassical marginalism and equilibrium theory has been lifted by the significant revisions of the new growth theorists, with the work of Krugman holding out perhaps the greater possibility of bridge-building from the mainstream to more heterodox positions. However, this is not to say that even Krugman's work fully escapes the restrictive assumptions of neoclassicism. What he has done is point to some important regularities such as production specialization as well as showing that new neoclassicism no longer needs to eschew imperfect information/competition and increasing returns assumptions to remain viable. Yet his work does not solve old problems but rather recasts them. This is particularly clear where the explanation for the growth and decline of spatial agglomerations and clusters is provided. For it is not really an explanation at all, rather an expression of a dynamic tension by means of which monopoly and competitive forms of the use of space work out, over unspecified time periods, an imputed internal logic of rivalry.

Still, for the moment it is the concerns of the various strands that make up the community of evolutionary economists that give greater satisfaction even though their solutions to problems posed may not yet have the elegance of Krugman's models. For, despite the clarity of the latter, it still remains a mystery in new growth theory as to why, for instance, despite the see-sawing logic of concentration versus deconcentration in economic development, divergence remains a predominant tendency regionally and globally, rather than the, albeit temporary, convergence that new growth theory predicts. This is largely because of the incapacity of neoclassicism to take account of institutional factors such as social capital, political regulation and economic innovation. The strength of the approach being laid out here is that not only is it interested in these dimensions of growth and change, it is committed to recognizing diversity. Neoclassical models normally fulfil the scientific criteria of generalizability and universality. But the world is made up of great diversity; regions with ostensibly similar histories differ dramatically in their growth trajectories, their capacity for lock-in or escape from it and their developmental 'absorptive capacity'. In what follows we shall explore why this is the case and what the answer tells us about how the minimum standards of developmental capacity can be raised.

3 Multi-level governance and the emergence of regional innovation network policy

Introduction

This chapter aims to explore important aspects of the governance dimension of Knowledge Economies. Of key importance in this is the clear implication of all decentralized industry policy that, on the one hand, what was called the 'circuitry' in Chapter 1 relates to regional and local nuances of industry's innovation requirements, and, on the other, that policy is similarly attuned. This raises a major problem for the majority of regional governance agencies and associations, e.g. business associations. To the extent the former are mainly public, they do not usually have the depth of experience and expertise necessary to discharge a variety of new and challenging functions, nor do they have sufficient budget to do more than influence rather than determine important functions like university research trajectories since national governments are normally the funding agencies for higher education. There is variation in this and states in the USA, for instance, have more leeway than most regions in, for example, Europe. It is a developing issue in light of the questions of 'proximity intellectual capital' posed by Knowledge Economies. For private actors in the regional innovation governance field, such as business associations, chambers of commerce, and industry forums involving social partnerships of trade unions and employers, may be more familiar with operating at national level historically and are thus no better equipped in general than their public counterparts. Probably one reason greater pressure is now put on universities to take the lead in regional economic development, apart from the obvious one that they are heavily implicated as sources of potential knowledge commercialization, is that they are among the few organizations in any given region with legitimate authority to speak knowledgeably on science, technology and, it is hence believed, innovation and the policies that support it. Thus they often become a key part of the regional governance structure in Knowledge Economies and regions with aspirations to move in that direction.

But because of the relative novelty of Knowledge Economies as defined in Chapter 1, and because of the substantial budgets involved in localities and regions functioning as fully sustainable Knowledge Economies, national and supranational (in Europe) governance of innovation promotion is essential. The most interesting, presently unique, case of this supranationally is the European Union, which has scientific, technological and innovation policy responsibilities and budgets. Indeed, as will be shown, it has taken on an important role in the context of major market failure in the European Union. That is, most European firms are a lot less innovative, especially in Knowledge Economy sectors than their North American counterparts, and outside a few of the main commercial cities there is a dearth of innovation market suppliers ranging from specialist innovation lawyers (e.g. patenting in specific sectors) to venture capitalists, although occasionally slight changes are discernible that show even hitherto weak and peripheral regional economies evolving linkages appropriate to Knowledge Economies compared to those inherited from their Old Economy structures. Northern Ireland is one interesting case of this as will be shown later in this chapter. It is instructive of the role an economic region of a nation–state can develop as a key actor in the governance of innovation developing Knowledge Economy capabilities.

The term 'region' is one of the most elastic in the social sciences. In past years it has frequently been deployed to denote collections of countries, viewed from a global perspective, rather like 'theatre' in military studies. Thus South-East Asia, the Middle East, central Europe and Latin America have all been referred to as regions, sharing some vague identity, common capacities or common problems, usually of an economic or security-related kind. This geopolitical view of regions is much promulgated in the International Affairs community and it has an intellectual pedigree dating back to nineteenth-century geographic thinking, often tied to geopolitical concerns with domestic and imperial security. The quest for topographically and culturally defined regions such as the true (Germanic) heart of Europe in *Mitteleuropa*, or Mackinder's identification of vast swathes of Eurasian territory as 'the pivot of history', testify to this tradition of concern with 'global regions' (see Agnew *et al.*, 1996). These 'georegions' are not the kind that are of interest here. Rather, the focus is upon the meso-level, or what European political scientists have come to call the 'third level', located below the supranational and the national, but above the local governmental strata.

The needs of this book make a focus on regions necessary for three reasons. First, our focus is on learning as an economic function that is of heightened importance in a knowledge-based era of business activity that many see as already having transcended the 'Industrial Age'. One key and evolving feature of this kind of economy is its interconnections with clusters as sources of external economies of scope and scale, operating in important ways at regional and sub-regional levels, the better to interact with georegional or global levels. Second, regions are relatively important levels of intervention capable of supporting activities of the kind described, while national, federal or supranational levels are too remote. Third, cluster policies are designed and funded by national governments often in partnership with regional agencies because the latter require taxation and business regulatory changes beyond their competences, to make them work. These might involve giving R&D tax credits or making changes in corporate tax rates related to share-option incentives, as was the case in the UK government's policy of incentivizing entrepreneurship and cluster-building in the late 1990s. But the regional level has responsibility for detailed design of specific clusters to support and the kinds of support to provide. So the need for 'joined-up government' also to the supranational level where appropriate, is crucial in this sphere.

The modernizing states of the nineteenth century had little time for regions as relatively autonomous power centres in their own right. Many were influenced by the Napoleonic model of the rational state with a strong administrative centre in the capital city and administrative representation in smaller departments or provinces, beneath which were democratically elected communes, none of which could or would threaten state hegemony. Apart from the New World where states often pre-dated the formation of the federal government, or the scale of territory involved made the establishment of federalized administration unavoidable, the evolution of regions within established state-systems, notably in Europe, has tended to be a phenomenon of the second half of the twentieth century. Two key processes are responsible for this. The first is an evolutionary process of *regionalization* or decentralization of powers from states to regions; the second is that associated with *regionalism*, the expression of demand for greater, or from some sections of a regional community, complete, autonomy from the state. These are complex processes that can be shown to interact with each other, although the impulse for regionalization is generally top-down while that for regionalism is usually ground-up.

Regionalization

This is the process by which, in general, a state devolves democratic power downwards by the establishment of Assemblies and allocates to them hitherto centralized administrative functions. The process is necessarily preceded by the drawing of regional political boundaries which encompass local or provincial tiers of government or administration but which so divide territory to minimize likely future instability for the state, either because of disproportionate economic, political or even cultural power being wielded by specific regions, or sometimes, by dividing cultural regions in some way. Thus the process can unite historically separate and politically or culturally dissimilar areas under a single regional administration, as in the cases of Emilia and Romagna in Italy, or Rhône and Alpes in France. Or it can divide culturally cohesive subsidiary 'nations' like Britanny, separated from its historic metropole of Nantes which was allocated to neighbouring but indistinct Pays de la Loire, or the Basque Country, separated from the Basque-speaking area of western Navarre in the Spanish regionalization of 1981.

Regionalization can be a relatively slow evolutionary process, as for example in post-war Italy where it took from 1949 to 1974 to implement the basic legislation from whence the scope of regional competences has gradually developed, though at two speeds due to the existence of the special (e.g. Sicily) and the ordinary (e.g. Tuscany) statute regions. The United Kingdom's largest country, England, is evolving through a process of slow and quiet regionalization from above. First, during the early post-war years administrative or statutory regions were established, mainly for purposes of collecting statistics but then as means of administering central government regional policy. Second, in April 1994 the pace quickened with the establishment of Integrated Regional Offices placing regional offices of central ministries for employment, environment, industry and transport under a single, regional administration. Third, in 1997 it was announced that there were to be regional economic development agencies for the English regions. They were set up in 1999 and managed by boards composed of nominated members. The quickening pace of English regionaliza-tion can probably best be understood in terms of demands from the European Union that to comply with Structural Funds requirements, there must be detailed, strategic regional plans for each qualifying region. Since all except the South East of England fall into that category to some degree, the central ministry responsible, the Department of Trade and Industry, was not able to meet the EU's new requirement for greater regional and local participation in the programme development process, let alone secure the appropriate range of regional co-funding. Nor was local government capable of fully meeting the new demands, not least because of the constraints imposed upon it by central government budgetary policy. The fact that the West Midlands integrated office had already established a new European Unit to take charge by June 1994 testi-fies to the motive force of European funding in the evolution of English regionalization (Burton and Smith, 1996).

There is a second, distinctive, kind of regionalization that, in historical terms, is much more swift in its implementation. The paradigm cases here are the federal states and, specifically, the European examples of Austria and Germany. As is well known, both were re-federalized in their re-foundation following defeat and occupation by the Allies in 1945. Hence, regionalization at speed can be said to be a predominantly *exogenous* impulse while regionalization at leisure is, to some extent, a more *endogenous* process. The exception to the latter part of that argument would be where endogenous regionalization occurs relatively rapidly following long periods of highly centralized control, as occurred at the extreme in post-Francoist Spain, less so in post-Gaullist France and in a mild way in post-Thatcherite Britain. But the Austrian and German cases are qualitatively different in respect of the *exogenous* imposition of new constitutions upon their countries which were both far-reaching and thorough-going in their effect. In consequence, those regionalizations are among the European instances of the highest degree of relative autonomy being enjoyed by sub-central political units, a position shared with the Belgian regions of Flanders and Wallonia, although they reached that status by a *regionalist* rather than *regionalized* route to be discussed below.

Endogenous regionalization tends, in general, to be associated with the devolution of the fewest powers and the lesser means to promote actions, either legislative or financial. Thus the French regional system is possibly the weakest of the regionalized arrangements. French regions have tax-raising powers but the source is mainly licence taxes for cars and registration fees on vehicle and property registration. Total regional expenditure rose from 6.9 billion francs in 1982 to 41.3 billion in 1992. The increase is largely explained by the transfer from central government of responsibility for education and training, with associated budgetary transfers for what transpired to be substantial capital expenditures. This process of regional involvement in central government expenditure arises from state–region planning contracts. However, in terms of financial resources, regional budgets amounted in 1992 to only some 8 per cent of total state expenditure in the regions, the vast bulk being incurred at local government level. Thus while the regional council representatives are engaged with the more powerful regional prefects in negotiations regarding priorities in France, the regions have little taxation or spending autonomy (Gilbert, 1994).

The Italian case is rather different, for while the regions have more taxation autonomy, they have little spending autonomy. Differences exist between the special and ordinary statute regions both in terms of the items covered and the weight of specific earmarked spending responsibilities in overall regional budgets. Thus, the special regions – all with some culturally distinctive base theoretically capable of expression in terms of political *regionalism* – spend proportionally more on welfare and economic development than the ordinary regions. The latter, although being the responsible authorities for raising motor vehicle and fuel taxes, as well as being the direct recipients of compulsory health service contributions, must spend nearly 70 per cent of their tax receipts upon the health service. Thus the ordinary regions spend only a little over 11 per cent of their budgets on economic development even though they are responsible for raising some 49 per cent of total regional expenditure through regional taxation (Desideri and Santantonio, 1996).

Exogenously regionalized *Länder* ostensibly have far more regional autonomy and competences with respect to taxation, spending, legislation and, importantly representation to, for example, the European Union. However, while there is a degree of truth to that statement, in reality the *Länder* of Germany and Austria are highly circumscribed. In Germany, major taxes are administered at *Land* level but the resulting total is divided between *Land* and federal levels, making a pool of 75 per cent of total federal and *Land* tax income. This is then allocated according to various vertical and horizontal equalization mechanisms. Agreements through committee negotiations also limit spending autonomy at regional level, something which is being further constrained by European Union competition rulings (e.g. on subsidies to declining industries). Thus even though German *Länder* have confirmed their right of representation by a representative of the upper tier of the federal government (*Bundesrat*) at appropriate Council of Ministers meetings, this does not necessarily enable them to bring influence upon Brussels to pursue regional innovation or economic development policies of the kind they would like autonomously to pursue (Sturm, 1998).

Austria, as a smaller federal country with perhaps stronger concertation practices between government, firms and labour is more centralized in its model of tax administration with the nine *Länder* receiving allocations from the centre in line with their social and economic needs and priorities. As in Germany, one way to develop regional economic policy at *Land* level in ways which fit well with EU aspirations for greater competitiveness and innovativeness has been the development of regional innovation policies. These are aimed at boosting the numbers of technology-intensive SMEs by encouraging networking and co-funding of innovation between federal and *Land* agencies, universities and industry. The case of Styria in lower Austria is an exemplar of this approach (Tödtling and Sedlacek, 1997). Austria's relatively recent accession to the EU means that new mechanisms for corporatist interest intermediation have had to be developed to protect the constitutional rights of *Land* governments. As Morass (1996) makes clear, this has resulted in the nomination of two *Land* governors as Austrian representatives on issues of regional concern at Intergovernmental Conferences and, more generally and as appropriate, at EU Council of Ministers meetings.

Thus the *Länder*, although circumscribed in their scope for action, unquestionably have substantially greater responsibilities, also budgets and degrees of representation than the regionalized bodies, at least in France and Italy, where the Member State conducts all negotiations with the EU on regional as well as other policy matters. The German and, to a lesser extent, Austrian *Länder* have been particularly active and innovative in promoting and part-funding initiatives to enhance their regional innovation capacity. Sturm (1998) proceeds to outline two different models for regional innovation promotion in Germany: regional centralism and regional decentralization. The former is practised in Baden-Württemberg where the regional government co-ordinates a growth coalition and supports it with policy instruments such as the Steinbeis Foundation for technology transfer from universities and polytechnics to SMEs, or the Ulm Science City financed by the *Land* government and large firms like Daimler and Siemens. This approach has been adopted in a number of new *Länder*, notably Thuringia, where Lothar Späth, former Minister-President of Baden-Württemberg is now active. The decentralized model is practised in North Rhine-Westphalia where regional 'conferences' have been promoted as mechanisms to induce more innovative potential into policies. This initiative 'from below' is being imitated in Saxony and Brandenburg amongst the new *Länder*. It is worth noting that French and Italian as well as Spanish regions such as Valencia have also recognized the merits of pursuing regional innovation policies, but their much smaller budgets and discretion inevitably circumscribe their capabilities in this regard (Cooke *et al.*, 2000).

Regionalism

Regionalism is a political process by means of which a culturally or linguistically distinctive part of a state that is territorially defined or definable as a sub-state region (or in its own terms 'nation'), mobilizes to demand greater autonomy over its own administration, its relations with the superordinate state and with supra-state bodies such as the European Union. Its fullest expression occurs when such a sub-state unit presses for and achieves political independence, as the Irish Republic did with respect to the United Kingdom. Other, lesser, expressions of regionalism would include Quebec *vis-à-vis* Canada, the Basque Country and Catalonia in Spain, Corsica and Brittany in France and Scotland and Wales in the UK. Because of the presence of some degree of historic cultural and linguistic identity, such regions usually possess a stronger institutional basis for regional autonomy than regions that have emerged through a process of regionalization. Such institutional ties rest upon shared cultural norms and modes of cultural reproduction that operate through associative practices rather than the state itself, though some recognition of distinctiveness by the state may be found through the tolerance or legitimization of organizations that differ from those of the state itself. Associative practices with political and economic development implications are explored further in Chapter 4.

Where such an institutional base exists, the quest is normally for the establishment of political organizations such as an administration and powers to make laws and raise funding for the pursuit of greater stability and security for the region in question. Importantly, such aspirations may have an economic dimension, particularly where past political or cultural discrimination by the state may also have been associated with economic weakness. Most regions in which regionalism tends to be pronounced have weaker economic indicators than the average for the state concerned, Catalonia being the main European exception since it is the economic powerhouse of Spain. While regionalism has its own evolutionary dynamics, such tend to be the sensitivities of states to the perceived dangers of not responding to regionalist pressures to some extent that such regions are normally the first to receive some compromise measures which go part-way to meeting the more ambitious demands of regionalist movements. Thus, in Spain the demands of the Basques and Catalans were early in being tackled in the post-Franco era; in Italy, the special statute regions such as Friuli–Venezia–Giulia, Trentino–Alto Adige, Sardinia and Sicily were established much earlier than the ordinary statute regions and in the UK, Scotland, Wales and Northern Ireland have had territorial Cabinet Ministers and supporting civil services with administratively devolved powers to vary spending patterns under a block-grant allocation, for a substantial length of time.

Regionalism, when limited aspects of demands have been met, can be the impulse for a wider strategy of regionalization, as has occurred, in effect, in all three of the states discussed above. Over time, a perception develops that devolution and, where it occurs, accompanying democratization, confer a certain advantage upon such territories. Others with comparable problems of peripherality or economic weakness but little or no devolution may press for a wider, symbolically fairer devolution of power. Of course, in the European Union, pressure by the EU for Member States to encourage greater consultation, participation and the formulation by regional actors of regional development plans as a prerequisite for the receipt of Structural Funds, has reinforced that trend, particularly in the UK but also in other unitary states such as Greece and Portugal in limited ways.

For those states that have established, either piecemeal or in a more thoroughgoing way, regional administrations in contexts where political regionalism has been present, the gamut of powers varies tremendously. The greatest autonomy has been achieved in Belgium where in 1993 a federal constitution was adopted giving the Flemish and Walloon regions, along with Brussels and the German-speaking enclave of Eupen, devolved administrations and assemblies with varying competences. The Spanish, Italian and French regionalizations gave extra status and competences to regionalist regions in the first two cases but not in the French case. In the United Kingdom, Scotland and Wales are not, in the strict sense, regions but nations, hence they are not, in administrative and political terms, beneath the state but rather part of it, as territorial Ministries of the UK government. We may briefly identify some of the key competences of selected cases to explore these distinctions and consider their meaning in relation to policy-making, especially with respect to financial autonomy, legislation and scope of representation.

In the Belgian system, budgetary and tax-raising affairs are a federal responsibility and, while there is a large measure of discretion in terms of expenditure at regional level, there are also equalization arrangements for negotiating fair allocations under specific broad budgetary headings. Regional governments have legislative competences on regional economic development, environment, transport, agriculture and export promotion. The communities, defined by linguistic identity, have responsibility for culture, education, health care, welfare and family policies. In Flanders the regional and community interests have been combined under the Flemish government and parliament, giving it wide-ranging competences. The regional governments have the right to represent the Belgian state at EU Council of Ministers meetings on matters of exclusive subnational competence such as culture, education, tourism and land use planning. Where competences are shared with the federal level but the regions have the dominant share, the delegation to the Council is led by a regional minister and assisted by a federal minister. Where the dominant share of competence is federal, the reverse applies. The former competences include industry and research policy, the latter matters concerning the Single European Market, public health and energy. A system of rotation of regional representatives ensures fairness of representation among the sub-national groupings. Thus, for some policy areas the Belgian regions act as second-level rather than third-level players in relation to the European Union within the overall Belgian legal framework. But, of course, to function effectively such action depends first on there being a consensus position throughout the state, otherwise representation is meaningless, especially where unanimity rules apply in European decision-making (Kerremans and Beyers, 1996).

Within the Spanish system, the Basque Country has the largest range of regional powers. On taxation, for example, and based on an historic tradition called *Frios*, the Basque provinces of Vizcaya, Guipuzcoa and Alava, rather than the Spanish inland revenue, collect most taxes (including VAT) since the post-Franco regional system has been established. Approximately 70 per cent are retained for the funding of expenditure within the region, the remainder being transferred to Madrid for defence, security and foreign affairs expenditure. Until 1997 tax rates had been broadly in line with those of Spain, but in negotiations with the minority Popular party government the Basque government gained responsibility for collecting and varying the rates of duties on petrol, tobacco and alcohol. Moreover, a new Spanish taxation regime gave regional governments powers to vary income tax rates by up to six percentage points, but the Basque government negotiated tax varying powers of up to twenty points. Legislation was also passed, somewhat controversially, to lower the Basque rate of standard corporation tax from 35 per cent to 32.5 per cent. Though these tax-reducing powers are not meant to make the Basque Country a tax haven, nevertheless they are directed at providing incentives to attract business to a region that has suffered from a negative image due to political turbulence. Thus the Basque Country has legislative power for the full range of domestic affairs, including responsibility for its own police force. However, under the strongly centralist Spanish government's position on foreign affairs, the Basque government does not yet enjoy the relatively modest levels of representation at Brussels that Belgian, German and Austrian regions have, though this too is under negotiation. Moreover, the Basques have highly developed informal links through Commission appointees and the presence of its well-staffed mission in Brussels, where the Basque perception of itself as a country not a region is symbolized by an extra gold star on their European Union flag (Marks *et al.*, 1996; White, 1997).

Finally, we may take the cases of Scotland and Wales within the UK system as countries with comparatively strong territorial ministries represented in the UK government but which, up to 1997, had neither democratic assemblies nor taxation powers, though Scottish legislative powers existed through the UK governmental system because Scotland, unlike Wales, possessed its own legal system. The 1997 Labour Government in the UK proposed a legislative assembly with tax-varying powers for Scotland but for Wales an assembly with no tax-varying and only secondary legislative powers. Until the outcome of referenda on these proposals, finance was allocated centrally through a block-grant system in which both countries received favourable extra disbursements based on a formula (the Barnett formula) in which allocations per capita are first worked out for England, these then being topped up for Scotland and Wales on the basis of economic indicators prevailing at the time the formula was established, when both economies were in the throes of severe industrial restructuring. Both administrations have powers to vary allocations under different headings within the block grant categories. Although both maintained missions in Brussels since 1992, representation to Brussels, especially the Council of Ministers, is always led by the appropriate UK government minister who may be accompanied by a Scottish or Welsh minister where an issue of special relevance is under discussion (e.g. the Welsh agriculture minister accompanied the UK minister in discussions concerning the BSE 'mad cow' crisis). This did not change significantly after assemblies were established in Northern Ireland, Scotland and Wales, though it is likely to be a subject for future negotiation within the UK.

Belgium, Spain and the UK are countries where regionalism has been comparatively pronounced and relative legislative and taxation autonomy within domestic state systems exists. Yet the meso-level does not thereby enjoy equivalent representation outside the Member States. The Belgian regions appear to have greater autonomy within their federal system than the Austrian or German *Länder*, but the strongest regions within the Spanish and UK systems do not. Thus for the foreseeable future, and despite claims of a tendency towards the emergence of a 'Europe of the Regions' and a 'hollowing-out' of the nation–state, much of the capacity and competence of regions for self-expression in political and economic affairs remains heavily contingent on the discretion of still-powerful Member States. However one new sphere in which there has been evidence of a coalescence of regional and European Union policy thinking concerns the emergence, especially among the regionalized but also the regionalist cases, of a realization of the importance for future economic development of adopting regional *innovation* policies. This has been one policy area with an evolving 'Europeanization' of policy and a degree of policy convergence among regions at different stages of economic development. As will be shown later in this chapter, these initiatives have occurred ahead of actions in this direction from Member States. Future growth of regional actions may be influenced by learning opportunities taken from observation of this phenomenon by both regional and European governance levels alike. For the present, innovative governance impulses seem more likely to emanate from these third and first levels of governance rather than the second, Member State level.

Such initiative-taking at the meso-level is not limited to the European Union, as David Osborne's (1988) book on America's 'laboratories of democracy' showed. The USA has a lengthy history of innovative state-level policies seeking to develop new economic practices. The Michigan Manufacturing Initiative and Pennsylvania's Ben Franklin Partnership to name two, both aimed to create more innovative state economies by learning from examples such as the 'Third Italy' and build clusters. Latterly, Massachusetts gubernatorial elections have been won by political leaders committed to building regional innovation capability by identifying and assisting development of knowledge-based clusters. North Carolina also has such a policy and Rosenfeld (1995; 2000) gives many similar accounts. Australia developed its networking and clustering initiatives after South Australia had adopted and implemented such an approach. While in Canada, a new period of reformist provincial administration in Ontario in the 1990s was accompanied by cluster-building learned from comparable models applied elsewhere, not least the Greater London Council. In the section of this chapter that follows, an extended account of the development of European Union innovation policy is presented and associated issues of multi-level governance of innovation are addressed.

Multi-level governance and innovation

The question of multi-level governance is most pronounced in the European Union because it is the only 'geo-region' with a supranational governmental system (i.e. elections, a parliament and administration, albeit without autonomous party-based decision powers). A specific sphere in which the issues this raises appear with particular salience concerns science and technology policy. Since these are fundamental to this book's concern with economic learning, innovation and clusters, multi-level governance of innovation policies is of specific interest. Since this is traditionally a national policy field in which the EU has been stimulating a regional response in some states while, apparently, curtailing it in others (see, for example, Grande, 1996), the discussion will be EU-centred. It is the most obvious case of a political system seeking to implement policies in active pursuit of learning economies, innovation, networking and clustering. This is because in numerous analyses of the EU's weaknesses *vis-à-vis* its competitors, innovation was highlighted as a crucial deficit in both business competitiveness and the quest for wider prosperity, cohesion and integration within the Union and one requiring a concerted and multi-level response (CEC, 1995). The regional level was seen as appropriate for this kind of policy because regional imbalance is such a pronounced feature of the EU space-economy.

Grote addressed this issue in advocating an EU regional policy for less favoured regions that 'would contribute to a more equal factor endowment of regions with social and organizational assets and resources (intangible software infrastructure) rather than merely dealing with the provision of traditional physical (tangible or hardware) infrastructure' (1993, p. 15).

There were attempts during the 1990s to take on board such injunctions, not only by programmes emanating from the Innovation directorate DG13, in the form of the Regional Innovation and Technology Transfer Strategies (RITTS) programme, but through joint action between DG13 and the Regional Policy directorate DG16 under the Regional Innovation Strategies (RIS) programme. The emergence of RIS signified the first step towards building 'soft' or intangible, network-form, infrastructures in less favoured regions to complement more typical past investments in transport and energy infrastructures. Yet the complementary action Grote called for, namely, an evaluation mechanism for all sectoral initiatives, to judge their overall cost-benefit in terms of direct and indirect impact upon less favoured regions, has not yet been forthcoming. This tells us that aid to less favoured regions continues to be seen as an economic cost not a benefit, despite the rhetoric

So, with respect to multi-level governance as between the EU and Member State levels, we see a process evolving, in the sphere of innovation policy, from an initial position which Grote (1993) describes as not even 'agency capture' but rather one of 'agency creation' by corporate interests in which multinationals, especially in ICT, pushed the EU into having a technology policy as long as they (the so-called 'Big Twelve') got the project-funding. One consequence of this was to create a new power for the European Commission,

Grande (1996) argues, as over time the room for manoeuvre by Member States and even European multinationals has become more circumscribed, especially where the social implications of technological change and the monopoly power of large firms are concerned. As Grande puts it:

> Even in the past implementing technology policy either as a nation alone or without the EU was problematic, in the future such action will be neither practical nor possible. And all attempts to limit the activities of the EU to peripheral areas and to secure the dominance of national policy must be viewed as having failed.
>
> (1996, p. 4)

In policy terms the national level had been circumnavigated in respect of innovation strategy, but so, by comparison with the early days, had the Big Twelve firms. This was due to the failure of both to come up with solutions to the problem of Europe's innovation deficit *vis-à-vis* the USA. This gave legitimacy to an alternative, pan-European Union, strategic approach, and enabled the Commission in the 1990s to take the initiative for developing strategy, albeit in consultation with Member States and the business lobby. But by the late 1990s, national governments woke up and began developing their own innovation approaches, often based, as in Germany and the UK on developing knowledge-based clusters.

One reason for this is the critique that became widespread when the latest 'Fifth Framework' policy for promoting science and technology was introduced in 1999. The two key dimensions to this critique reveal elements of 'lock-in' to earlier Commission thinking, which most science policy experts think is now anachronistic. First, the model of innovation remained principally a linear, 'technology-push' approach. Second, the approach accordingly privileged research over application and therefore underplayed social and the regional elements of what most would now see as not a linear but an interactive, user-driven model of innovation processes. These dimensions were interconnected since both suggested the need to grasp a more systemic concept of innovation, and this the interests involved in the policy formation process failed to do. Thus, while EU policy stimulation of interaction between researchers in the different Member States improved relationships in the horizontal dimension, the impact of this vertically, in relation to the competitiveness of industry and the expansion of employment at national and regional levels was limited. The problem lay in the difficulty European firms displayed in turning good basic R&D into commercial innovations. This is a result of the lack of effective innovation (and technology) policy to promote applications based on the findings of fundamental research. Meanwhile, the tacit-knowledge exchange between users and producers of knowledge and the various intermediaries such as technology-transfer agencies, consultants and specialist firms involved in interactive innovation in geographical co-location was being shown daily to be at the heart of US success as an innovative economy.

The regional element of the multi-level governance process was overshadowed by the expressed interests of industry, the technology-intensive northern Member States and the Commission. This is echoed the muted role identified for the regions in the Commission's *Green Paper on Innovation* (CEC, 1995). There, in the analytical section of the report, regions are linked to SMEs but the whole of a brief two-page summary of their importance to innovation is devoted to an account of SME characteristics with no mention of the role of regions in innovation support. Later, however, in outlining thirteen 'routes of action' for the future, one of these refers to the regional dimension of innovation, hardly in the most ringing terms: 'The local or regional level is in fact the best level for contacting enterprises and providing them with the necessary support for the external skills they need (resources in terms of manpower, technology, management and finance)' (ibid., p. 57).

But further detail is added, listing what 'contacting' was meant to imply: the fostering of inter-firm cooperation; encouraging technology transfer; forging information and communication networks; and reinforcing university–industry co-operation, while the EU's role was to strengthen inter-regional cooperation; to support interactive innovation; to co-fund regional innovation strategies; to strengthen Business Innovation Centres; and to introduce training for innovation policy personnel. Many such functions have been mainstreamed into eligible expenditures capable of being funded by the European Regional Development Fund and other Structural Funds.

This does not yet presage Jeffery's (1996) notion of a farewell to the Third Level. He saw the ending of the strong impulse coming from the German *Länder* for the establishment of strong representation in EU governance at the third or regional level, best expressed in the pressure for the establishment of the Committee of the Regions. Here, the argument is that the *Länder* have a high capability for influencing EU policy priorities through their constitutional involvement in German Member State policy formation towards Europe, and they have been influential in pushing for a strong Committee of the Regions. This, the optimists believed, might one day play the role in the EU that the *Bundesrat* does in Germany, i.e. as the second chamber of the European Parliament.

The mobilization of regional capabilities for innovation

In previous sections of this chapter it has been argued that processes of regionalization and regionalism produce rather different outcomes in relation to the mobilization by regions of competences to become significant policy actors, within their Member State or above it to the supra-state body to which they belong. In principle, regionalization, as implemented in federal systems such as Austria and Germany, produces a constitutionally equal sub-state playing field, though the advantage taken of competences assigned will vary according to the astuteness and capabilities of regional actors. Where regionalization produces uneven competences it is because of regionalism, as is clear in the cases of Italy and Spain where cultural and, often, linguistic distinctiveness accompanies special regional status and greater regional powers. Where only regionalism operates as the impulse for the demand for greater representativeness, as traditionally in the UK, there is accordingly greater unevenness in the acquisition of regionally representative administrative and political organizations. However, it is worth noting that in either case there can be an imitation or aspiration effect in that unrepresented regions may seek to gain some, at least, of the powers enjoyed by those of special status. The case of Belgium is also instructive as being the product of both regionalization and regionalist impulses, and it has developed the most far-reaching and relatively equal forms of regional representativeness both internally and externally, at least in Europe.

The question of regional *receptivity* to new demands for innovation and enhanced learning capability is important, particularly the exploration of policy mobilization and the conditions for high general receptivity, learning and 'absorptive capacity' by regions. The first focus will be upon regional mobilization and the second upon regional receptivity. In this, the relationships between regionalization and regionalism will again come to the fore. Whereas regionalism has tended to be presented as a basis for mobilization deriving primarily from solidarities of an ultimately ethnocentric kind, such as distinctive linguistic–cultural attributes, it will now be necessary to take account of what Harvie (1994) refers to as 'bourgois regionalism' where hitherto regionalized but culturally relatively undifferentiated regions such as Baden-Württemberg, Rhône-Alpes and Lombardy have begun to exert their civic cultural bases to demand enhanced economic autonomy from their states. This will prepare the way for an assessment of the findings of research, such as that of Putnam (1993) on civic culture and 'social capital' in Chapter 4, particularly among non-regionalist regions. These point to competence and capacity being intimately related to the capability of regional communities to capitalize upon 'associative' strengths for purposes of regional political and economic development (see also, Cooke and Morgan, 1998).

Regional *mobilization* in the European Union is a relatively new concern that has exercised political scientists interested in the evolution of multi-level governance (see for example Mazey, 1995; Hooghe, 1996; Jeffery, 1996). Setting this within a multi-level governance framework, Marks *et al.* (1996) approach the question in terms of whether the underlying logic of mobilization is fundamentally related to economic or political resources. Like others, they take the establishment of regional missions in Brussels as a key indicator of mobilization and a sign of the evolution of multi-level governance 'from below'. But their interest is in why some regions have done this and others not, and how to explain the emerging patterns. One explanation would be that economically better-resourced regions could be expected to establish missions on grounds of affordability and a high valuation of likely returns on information gained from proximity to decision centres – a public choice impulse. The alternative explanation is that such representation is sought for political reasons by regional authorities at odds with or perceiving under-representation from their Member

States, associated with their distinctive regional, cultural identity within their state – in other words, a regionalist impulse. Five hypotheses are then advanced for statistical testing against appropriate data: resource pull (or rent-seeking behaviour); resource push (or affordability); associative culture (social capital); competence (valuation of information); and regional distinctiveness.

The conclusions of the study are that resource pull, measured in terms of the regional distribution of European Union Structural Funds, gains little support from the data and that the regions that gain most, i.e. Portugal, Greece, Ireland and Southern Italy have not a single regional mission among them. The explanation for this could be that such is the relative influence of the EU and Member States over Structural Fund budgets that there is insufficient scope for a regional office to be worthwhile. The resource push hypothesis based on data on regional revenues also gains no confirmation since economically less favoured regions such as Extremadura or Murcia in Spain retain missions while there is no regional representation, for example, from rich regions in Denmark and the Netherlands. The associative culture hypothesis does gain statistical support but the statistical measure deployed is far from adequate. The competence or informational exchange hypothesis, measured by a score given on an index of regional political autonomy, is strongly and significantly correlated with regional representation. This is confirmed by interview evidence that information transmission between regional administrations and the Commission and back is their prime function. Thus regions such as those of Belgium and the German *Länder* are all represented while those in unitary states like Portugal, Ireland and Greece are not. Finally, the regional distinctiveness hypothesis, that regionalism explains regional representation, gains strong statistical support and sustains that support more strongly, against a battery of tests, than the other hypotheses.

Hence, on the basis of this brief analysis of Marks *et al.*'s (1996) extensive findings, there is evidence that regionalism is the strongest motivator for regions to mobilize to the extent of establishing a regional mission to the European Union. Thus in the UK, Scotland, Northern Ireland, Wales and Cornwall have representation and are regionally distinctive, whereas English representation is confined to the counties of Kent, Lancashire and Essex (shared with Picardy). In Spain, the Basque Country, Catalonia, Galicia, Valencia and the Canary Islands are represented while only Andalucia and Madrid join the aforementioned Extremadura and Murcia missions as non-ethnically or linguistically distinctive regions. Marks *et al.* (1996) conclude that, in general terms, the greater the overlap between state and regional competences – as found in *Länder*, the Belgian case, and regionalist cases – the greater the likelihood of regional mobilization for purposes of representation. Those regions with the least overlap, though in many cases, the greatest economic benefits from the EU, are the least likely to mobilize. The answer to the question of why regions mobilize to the extent of establishing missions to the European Union in Brussels, is that it is a primarily a political rather than economic logic that drives the process. However, this finding needs to be nuanced by that of Jeffery

(1996) that missions are strongly associated with 'grantsmanship' i.e. improving opportunities for accessing grant-aid from Brussels, mainly through better, swifter information access or, to a lesser extent, influencing policy formation. Moreover, Mazey (1995) points to the scarcely coincidental establishment of most offices in the run-up to the implementation of the 1992 Single Market legislation. Thus, it may be concluded that while a regionalist or *Länder*-based political logic explains the kind of regions that have mobilized earliest, a deeper economic anxiety underlies that political logic.

The question of *receptivity* as the capability to develop policy that may assist the development of innovation and a learning economy involving, for instance, cluster-building requires less of an external and more of an internal focus on regional assets. Starting from a definition of 'region' as a territory less than its sovereign state, possessing distinctive supra-local administrative, cultural or political power, differentiating it from its state and other regions, it can be hypothesized that strong regions are characterized by sets of characteristics that operate at three distinct but interacting levels (see Cooke *et al.*, 1997; 1998).

The first of these is the institutional level, meaning the prevailing norms, habits, customs and mentalities in the regional community at large. In the more accomplished and capable regions this is typically cooperative rather than simply competitive, but competitive elements may clearly also be present. By extension there is an associative culture rather than an individualistic one in the sense that actors will expect to operate through collaborative mechanisms such as clubs, societies, partnerships and associations by virtue of which social capital is exploited in the civic sphere. A learning disposition will be pronounced as a cultural element rather than an introverted 'not invented here' syndrome. Thus formal and informal 'benchmarking', measuring performance against internal expectations or goals, and the practices of external peers, is a pervasive cultural feature. Implied here is a willingness for change to be accepted rather than resisted and for public–private consensus to prevail over dissensus. Thus, a regional culture of 'embeddedness' rather than 'disembeddedness' prevails (Granovetter, 1985).

The second level, and one on which more will be said in the following chapter, is organizational and refers to the practices of firms. Organizations are the media, which convey institutional norms, and firms are highly important actors in respect of the innovativeness and competitiveness of regional economies. In accomplished regions, firms will have evolved trustful rather than antagonistic labour relations, have created an environment of workplace cooperation rather than division and a worker welfare orientation as distinct from a 'sweatshop' mentality. This will further be characterized by the adoption of a 'mentoring' approach to workplace learning rather than a 'sink-or-swim' attitude towards skills development. Such firms will be comfortable with the externalization of production through supply chains based on long-term, preferred-supplier contracts as distinct from internalizing as much production as possible. And firms will tend to be innovative rather than simply adaptive to market trends.

The third level is also organizational, but refers to the sphere of policy-making within and between administrative and other regional spheres of activity. Thus, policy-making will be inclusive rather than exclusive, based on policy networks rather than on internal policy hierarchies alone. Policy will be informed by monitoring and the valuation of early access to internally or externally generated knowledge and information on regional performance, rather than reactive to problems of an unanticipated nature. There will be considerable delegation rather than centralization of functions, with the notion of responsible self-management informing the 'letting-go' of competences that may be more appropriately exercised by subsidiary, even private, bodies. Consultation will prevail over authorization in the policy-making process and networking will be widely engaged in, rather than policy-making being based on a tendency for administration to take a 'stand-alone' attitude.

Naturally, the extent of competences such as the presence of autonomous taxing and spending capacity and the availability of significant public and regionalized private financing capacities are extremely important to likely regional performance. The ability to influence the programming of strategic infrastructure is also key, whether of the hard kind such as roads, utilities and telecommunication or softer kinds such as the promotion of strong university–industry linkages or robust vocational education and training systems. The competence and capability for regional mediation and promotion, whether to put together financial packages or market the region to inward investors, are also important. And, finally, the capability to formulate creative strategies, as, for example, with Regional Innovation Strategies, rather than piecemeal, opportunistic projects, will be the hallmark of the active, accomplished regional policy-formation process.

It can be inferred that some of the regional cases mentioned in this chapter approach many of the features described in this account of the accomplished regional governance structure. But none indisputably possesses all the attributes listed. Moreover, the vast majority of regions, even in the European Union, let alone outside it, display few, if any of them. However, in an evolutionary development process in which the positive attributes increasingly operate as a kind of evolutionary 'selection mechanism' associated with social and economic accomplishment, more and more regional bodies will seek, within the confines of their own historic path dependencies and position within their own state-system, to emulate if not imitate successful regional models. In the multi-level governance system of the European Union, that supra-state body itself is an important influence upon the aspiration of regions and Member States to improve integration and cohesion within the Union through developing these characteristics of embeddedness at the regional level. The most obvious examples are the cluster and innovation-orientated strategic plans of the new Regional Development Agencies in England. These have been incentivized by the UK government through 'Cluster Challenge' to bid for funding to help establish new firm incubators as 'seed crystals' for their proposed clusters. These were generically seen to be biased towards 'new economy' rather than 'Industrial Age' clusters to fit the 'knowledge-driven economy' emphasis of the national government's industry policy of the same name.

If we seek to understand the way multi-level governance works in enabling Knowledge Economies to evolve, we can of course, consult cases of successful development like Silicon Valley. It exists in a regionalized governance system defined by the federal constitutional arrangements of the USA. There is distinctiveness within the Californian innovative enterprise support system because it is heavily dependent on the market rather than public support. Federal public research funding and state funding for public universities sit side-by-side with private research and training systems, set within the US federal funding and regulatory regime. Multi-level governance proceeds but government is mainly a supplier of basic research funds in, for example, health, defence, space exploration and some other fields. At the 1985 peak prime defence contracts were worth $35 billion to the California economy, of which a substantial portion was for research. But as many accounts such as Saxenian (1994) and Florida and Kenney (1990), let alone Zook (2000) show, Silicon Valley basks with the protection of a massive venture capital 'parasol' and a platform of abundant federal research and prime defence contract funding.

As Figure 3.1 shows in sketch outline, the role of the state legislature and municipalities pales into the background, though not into insignificance for lobbying, the all-important means by which those research dollars are diverted or, as Henton *et al.* (1997) show, as partners in a future-oriented forum like Joint Venture Silicon Valley. But apart from administration and tax breaks for research investment, the main role of the state, according to the Henton account was through the governor's Office of Research and Planning calling a

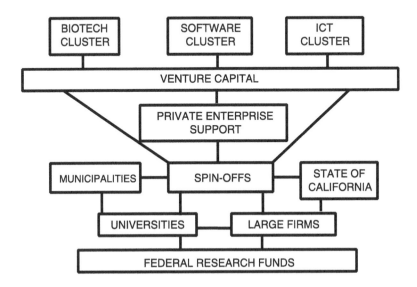

Figure 3.1 Simplified version of Joint Venture Silicon Valley capital 'parasol'

forum to report on *California 2010* and declaring a 'California Compact' to guide the state forward, dealing with broad economic development, traffic and land use issues. This led to Joint Venture Silicon Valley, a partnership between civic entrepreneurs from public and private sectors. But compared with the impact of other actors shown in Figure 3.1 it is clear that the public sector is in the background except for the massive infusions of R&D funding and the funding of public universities. Hence, there is a strong funding but weak policy role for the public part of the multi-level governance system operating upon Silicon Valley. Nevertheless, without it there would not be a Knowledge Economy upon which knowledge could act productively, creating major demand for private investment that takes the commercialization and innovation processes forward.

The federal funds platform is substantial but is dwarfed by the scale of venture capital available to support incubation and late-stage (initial public offering, IPO) investment in new businesses, the carriers of innovation from corporate or university laboratory benches to commercialization in local, national and global markets. Both directly through cross-holdings of equity stakes in spin-off and start-up firms, and indirectly through geographical prox-imity, venture capitalists are prime movers in the formation of sub-sectoral clusters such as biotechnology, software and ICT (including new media). Thus the Knowledge Economy has given rise to all industries associated with the New Economy, where knowledge acts on itself, mediated by humans and machines in techno-economic networks (Callon, 1991), to create productivity and enhanced economic value. That this is also a Digital Economy with digital value chains at its heart is a product of the ubiquity of this master technology within the trans-actional relationships of the clusters and the technological 'circuitry' being deployed to produce value.

The system can be compared with that operating in a more weakly and partially connected way in Northern Ireland, a more regionalist kind of governance system. It has an economy that, although part of an advanced national innovation system with the full panoply of multi-level governance, has been troubled by polit-ical strife and is particularly dependent on a 'pyramid' base of public enterprise and innovation support. This narrows towards the firm level where a majority of indigenous firms are non-innovative and where innovation comes from three small groups of firms: multinationals, indigenous innovators in traditional indus-tries, and Knowledge or New Economy start-ups. As Figure 3.2 shows, various external sources in government and particular markets are important in providing the platform for regional economic development. Figure 3.2 is illustrative of the embryonic Northern Ireland innovation system. It is not yet a fully functioning system but has certain emergent sub-systemic qualities, as has been noted. In general, it aspires to become better networked ('joined up') throughout and to strengthen the innovative capabilities of the different groupings of firms identi-fied, particularly indigenous non-innovators (Cooke *et al.*, 2001).

The governmental platform is ill-attuned to the exigencies of the New Economy, with some exceptions. To build strength in that dimension, the

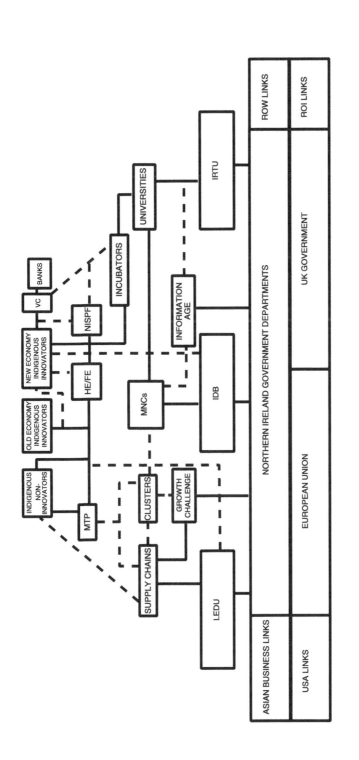

Figure 3.2 Northern Ireland's partial innovation system pyramid

institutions in the middle horizontal and upper right quadrant need strengthening and integrating since these are the ones helping replace Old Economy industry. Something of the order of a hundred smaller start-ups in Internet and telecommunications software are thought by the local economic development unit (LEDU) to exist in Northern Ireland, and foreign investors like Nortel Networks and Fujitsu are candidate customers for such technologies as optical networking. But the industry could benefit from better horizontal interaction and collective learning to enhance its own sustainability in global markets. Only if the New Economy indigenous and multinational firms find reason to interact with university research for innovation on a growing basis would it be possible to say that Northern Ireland had developed systemic regional innovation capabilities.

Nevertheless new intermediaries have, at the turn of the millennium, begun to put out 'feelers'. Supported by the Peace Process and special funding to boost economic activity, some from countries like Australia and New Zealand, some from the USA and the European Union, the rest from the UK government and special Knowledge Economy initiatives in Northern Ireland, led to the halting formation of linkages between strong pre-existing organizations and new nodal, networking bodies. Growth Challenge for cluster promotion, Information Age for ICT promotion, Manufacturing Technology Partnership for technology transfer, and the Northern Ireland Science Park Foundation are new intermediaries in the system building process. Much of the established enterprise support infrastructure was charged with attracting foreign investment through the Industrial Development Board (IDB), supporting SMEs (LEDU), and allocating industrial technological research grants through the Industrial Research and Technology Unit (IRTU). In 2001 these were amalgamated into a single, supposedly leaner enterprise support organization. But a set of interesting developments in the direction of the New Economy arose prior to this from the work of universities establishing incubators, helped by EU funding, where start-ups and spin-off firms could be supported first by IRTU grants, later by UK government Science Entrepreneurship and 'University Challenge' funding to assist commercialization with local venture capital to progress firms towards IPO positions through equity investments. The latter provide strong management support and have firms in similar sectors being encouraged to inter-trade, an important step on the route to forming clusters. But all levels of governance have been involved in stimulating these shifts.

Conclusion

In this chapter an attempt has been made, through a focus on the emergence of regions as evolving members of multi-level governance systems, especially in the European Union, to delineate the key characteristics, competences and mobilization processes by which the complexities of multi-level policy processes occur and themselves evolve. By reference to two key processes: regionalization and regionalism, it was shown that the former produces a more even distribution of policy competences within a given state-system, but the latter produces, at the peak, the more salient powers and policy capabilities. Taking as the focus the emergence of the new policy field of regional innovation policy, it could be seen how it evolved from being mainly a policy controlled and influenced by large industry and Member States to one in which there was an increasingly Europeanized model of regional development policy embedded within it. This has involved limited recognition by the European Commission of the critique that traditional technology policy reinforced regional disparities, thus under-mining its master goals of cohesion and integration. It also involved the formation of bridges between different divisions of the European Commission and the opening of communication linkages to the regions, albeit mediated by the Member States to varying degrees.

The receptivity of regions to this process was shown to be a learning process involving acquisition of knowledge and information of consequence to the development of regional innovation policy in both the horizontal and vertical dimensions. With respect to the latter, the mobilization by some regions to enhance their capabilities by establishing regional missions in Brussels can easily be seen to be consistent with the earlier analysis of regionalization and region-alism processes. The more accomplished regions were the administratively strong *Länder* and the more culturally embedded regions with pronounced regionalist characteristics. For political and deeply economic reasons these are the leading edge in terms of representation within the European Union system of multi-level governance. An analysis of exemplary characteristics of accom-plished regions showed that a high degree of administrative competence and regional embeddedness is of crucial importance to mobilization and capability to achieve govern innovation policy, an exemplar new regional function. The development of embeddedness is itself an evolutionary process, important elements of which can be learned, though not necessarily imitated directly. Moreover, as will be shown in Chapter 4, it needs to be complemented by 'autonomy' of a different kind than that normally associated with political self-governance for sustained growth to be practicable.

In two contrasting cases, one from the world's most accomplished New Economy location, Silicon Valley, the other from a less-favoured, peripheral region of Europe, Northern Ireland, it was demonstrated how public funding has a strong role to play at the base or as a platform in both. In the USA, such funding is mainly federal, in the UK it is mainly national. But for development of the innovation capabilities that take regions to a more pronounced position as Knowledge Economies, the role of private investors and the management capabilities they bring through private social capital are crucial. This shows how much of a sea-change in policy thinking is required as regions move from the Old to the New Economy, or fail to. Much more of that effort rests on building entrepreneurship for innovation. Small and medium-sized enterprises need management support more than the subventions to capital investment that old-style regional policy was designed to provide for inward investors. But public support funding is not in harmony with that, any more than it is expert in innovation management in technology-intensive businesses. Hence, if start-ups can find the New Economy intermediaries that can meet their demand, which is mostly of the kind normally met by private investors and innovation support firms, they will often pay them for their services either in fees or equity. The tragedy for such firms in many regions is that they do not find assistance of that kind from the private sector due to classic market failure reasons, i.e. risk-aversion by private capital, so perforce innovation in such settings is dependent mostly on public elements of multi-level governance. Better than nothing, but as has been suggested, ill-attuned to the task at hand except as fundamental supplier of the research funds on which venture capital and successful start-up businesses ultimately thrive.

4   Learning, trust
and social capital

Introduction

The idea of learning economies has evolved in connection with the development of the knowledge society. The knowledge society or knowledge-based economy is the latest stage of the process by which, first, information became a defining feature of the resource landscape of late twentieth-century advanced economies and, second, the selective exploitation of appropriate and useful information became organized as knowledge-based industry. Probably the paradigmatic instances of knowledge-based industries selectively appropriating information as a resource to be exploited are industries such as financial services, information technologies, biotechnology and biosciences and cultural industries. In each of these, highly specialist knowledge is the key resource from which innovations flow, products, processes or services are developed and significant returns on investments made can be earned.

The kinds of knowledge for these fields of exploitation are different, yet their wellsprings lie in the still mysterious bedrock of creativity. The creativity to imagine the molecular structure of DNA is different in kind but not necessarily in magnitude from that of Cézanne imagining the structure of landscape in terms of cylinders, cubes and cones. The evolution of the former discovery into modern genetic engineering and the latter into cubism, abstract art and beyond, to the dissections and formaldehyde of Damien Hirst, suggest a deeper industrial-age drive to understand and unlock the mechanisms of life through the metaphors of science, with the body, animal or human, the last visible resource to be exploited. And as an apparently infinitely renewable resource, the potential value of such exploitation is hard to over-estimate.

Yet these values pale into insignificance when set against those generated by innovations in, for example, software as promoted by Microsoft and others, or bond speculation and other finance industry practices that emerged in the 1980s. How do creative processes evolve in settings such as these? In his account of the development of Microsoft's multimedia encyclopedia *Encarta* and its successors, Moody (1995) describes the corporate campus where the creative work for such a venture is done as follows:

> The atmosphere on the campus is one of unrelenting anxiety and constant improvisation. Microsoft is ceaselessly assembling and reassembling its recruits into small teams of engineers, designers and editors, turning them loose to dream unsupervised then eventually calling them on the carpet and demanding that they produce something.
>
> (1995, p. 4)

The picture is one of chaos, absence of leadership, highly-focused engineers incapable of communicating in a civilized way with creative personnel brought in to design the graphics, or project managers brought in to try to get at least some work completed to at least one deadline. Yet the chaos masks a rather rational production plan and process involving the following:

1   Designers, editors and marketing personnel sketch out a rough plan for a product.
2   Designers construct a small product prototype on a computer for demonstration to developers, managers and users.
3   Design team accumulates content and works on specific features, ultimately writing full, detailed specifications for the project.
4   Documentation handed over to the developers who write the code which meets the design team's specifications.
5   Designers and developers iterate on solving problems in writing code for meeting specifications.
6   Designers move on to next project leaving developers to complete the project.

This is a highly team-based approach, in which a mixture of necessary skills is brought physically together. Some are employees of Microsoft, many are free-lancers, teams coalesce and diverge, specific skill profiles predominate, then recede according to the expertise required. The project is the focus, the campus is the location and the interaction between diverse personnel is the creative process. Yet it is also a process riven with creative and not so creative conflict. The principal conflict is between the dreamers and the doers, the designers and the developers. The developers constantly see the demands of the designers as unfulfillable while the designers see the developers as obstructive 'nerds'. Interestingly, though, these conflicts, serious as they can be, do not cause or reside in lack of trust or absence of faith in reputations, except in occasional instances. Propinquity allows for the evolution of recognition of worth, as well as its occasional opposite, but creativity seems to require a degree of conflict and stress, on occasions, that belies the element of mutual esteem. The setting is thus one of a constant tension between cooperation and conflict, collaboration and competition.

   In what follows, we will explore further the processes that characterize learning under conditions where cognitive dissonance, mutual misunderstanding and communication breakdown are highly likely. Both within and between economic organizations there are points at which the ideals of smooth information flow, clear understanding of roles, and regular communication of necessary knowledge to solve problems or exploit new opportunities are met less than optimally. If internalized learning within an organization is prone to interruption, how much more so is that which necessarily occurs between organizations in the externalized economy of supply-chains, sub-contracting and the like which now typifies so much of the contemporary transactions activity of the modern firm? Creativity and innovation are highly dependent upon learning and learning is a socially interactive process, albeit one which may also involve socially produced inanimate objects such as encyclopaedias, whether in book or computerized form.

Social learning

One of the widest, most recent surveys of the centrality of evolution to social life, and one that is even accepted as occupying a complementary, yet upbeat, position to his own rather doleful conclusions, concerning the 'selfish gene' (Dawkins, 1976) is that of Ridley (1997). He concludes that the origins of virtue lie in trust. Accepting the 'selfish genes' thesis that we exist as vehicles seeking to ensure the immortality of our genes, Ridley recognizes that to do so we operate in ways that are social, trustworthy and cooperative. The predisposition to learn these attributes and assess their superiority over self-interested opportunism is what, he argues, distinguishes humans as a species. 'Instinctive cooperativeness ... sets us apart from other animals' (1997, p. 249). In case the impression is given that he is irremediably naïve, he also points out that social relationships fragment and societies are prone to dissent, division, and varieties of anti-social behaviour, not to mention war. Certainly, competitive instincts predominate in social life and dominant ideology in advanced economies often presents this as a highly desirable social and individual characteristic.

It is often thought quite unexceptional to place cooperation and competition in opposition to each other, yet the resulting paradox has a debilitating effect on our efforts to decide the important from the less important features of social life. In contrast, it was once memorably asserted that the opposite of competition is not cooperation but monopoly (Williams, 1983). Where does that lead us? All three concepts rest on a notion of trust; indeed monopoly, in the sense of cartels, is even described as such, and anti-trust legislation exists to regulate the too-powerful exercise of it. Williams also reminds us of the institutional bases of the modern economy, in which most aggressive competitors must place their trust: the capital market, property ownership, joint-stock companies, free trade, international monetary systems, labour markets, etc. Cooperative behaviour is less obscurely rooted in the trust that partners will not breach faith or contract but will reliably be 'as good as their word' in the exercise of actions to complete an agreed project or meet a shared goal. Many authors have sought to explain the role that trust plays in social life. Three definitions of the concept published in the 1990s emphasize elements relating to confirmation of expectations, confidence and fulfilment of obligations. Thus Sako (1992) defines it as 'an expectation that your trading partner will behave in a predictable and mutually acceptable manner', while Sabel (1992) defines trust as 'the mutual confidence that no party to an exchange will exploit the other's vulnerability' and Ganesan (1994) as a 'belief or expectation based on the partners' expertise, reliability and intentionality'. Each of three definitions is inspired by thinking of trust in economic terms. Sako and Ganesan focus on customer–supplier relations, Sabel on networks of cooperating firms involved in the pursuit of a particular set of project objectives.

The interest in trust in relation to supply chains is understandable for precisely the reasons mentioned in the introduction to this chapter. Control of external relationships between independent actors is impossible if trust is not understood and learned. Sabel speaks of 'studied trust' almost as a technical instrument that has to be added to the manager or policy-maker's battery of decision tools. Stevens (1989) refers to a supply chain as a system of actors and facilities (customers, suppliers, logistics, production facilities) linked by the forward flow of materials and the feedback of information. Information on performance of the system, especially in relation to quality, flow disturbances due to equipment or task-implementation faults, and decision requirements makes the system a permanent arena for learning processes. Thus monitoring is a necessary part of the process whereby the intentions or objectives of the project in hand (e.g. Microsoft's *Encarta*) are compared with actual perfor- mance by means of indicators (numbers of rejects or software bugs, for example). Statistical information can enable learning to occur with respect to specific kinds and locations of good or poor performance. Responding to the results of the exercise of 'learning capital' involves decision procedures linked to decision points where the knowledge of monitored underperformance can be assimilated and action taken to remedy it. Decision actors and information suppliers may have opportunistic reasons for concealing information that can harm their reputation, in which case the supply-chain system cannot function. Thus it is clear that trust, in this context means openness, honesty, and willing- ness to volunteer reputation-threatening knowledge as well as the more general elements of confidence, reliability and predictability.

The notion of trust as openness is interesting because, on the one hand, it can be expressed as an ideal version of a true democracy as in the title of Popper's (1965) book *The Open Society and its Enemies* which was a critique of socialist societies such as the Soviet Union. On the other hand, it is a common practice of Japanese firms engaged in forging supply-chain agreements with their sub-contractors. Typically, firms such as Sony require their suppliers to 'open the books' to show them in every detail both how they design and cost the product or service they are selling. This is typically described as a learning process for both the customer and the supplier. The supplier will be one of many, so knowledge transfer of design improvements or cost savings on mate- rials or labour processes can be effected from one supplier to another through the customer, who obviously, under such circumstances, is in an extremely powerful position in relation to the suppliers. For the customer, learning best practice from suppliers leads to a better, cheaper product or service. It also reduces uncertainty by maximizing information flow. But why would any sane supplier give away its secrets in exchange for a contract that might be short-lived? In the case of Sony, the argument is that they do not make, they only buy, so they themselves have no interest in copying the technologies they are learning about, merely transmitting best practice throughout their supply-chain system.

But for the wary supplier this is still a risky business. One supplier interviewed by this author (see Cooke *et al.*, 1993; Cooke and Morgan, 1998) had been approached by such a customer and had been assured that his excellent company's microelectronics product, which was produced in Singapore, was of great interest but that to gain entry to the sub-contracting process, the supplier would have to 'open the books'. This, the chief executive of the company decided not to do. His argument, on the basis of wide reading and deep thinking about Japanese business culture, was that the tradition of *keiretsu*, a supply-chain system in Japan, typically consists of a key customer and a family-like set of suppliers. The family-like dimension extends to the normal situation in which the customer owns a substantial equity share in a given supplier. For this reason, the customer has an extra interest in ensuring the supplier is performing well since the customer benefits from improved share performance by the supplier. Our Singapore interlocutor's view was that the high-trust relationships of the family could not be reproduced outside the family. This was illustrated from his own experience when he gave the example of discovering that although the customer firm seeking his services was not a competitor, one of its 'family', jointly owned by three large Japanese firms, was. 'If I had opened the books to that company, our hard-won know-how would have disappeared into their *keiretsu* like a snake with a mouse,' was his concluding remark.

Yet, in many cases, firms do 'open the books' to customers. In another case reported in Cooke *et al.* (1993) the entrepreneur in question initially surprised even himself by agreeing to open up to Japanese firms. The owner-manager of the supplier firm in question professed a philosophy 'based on trust above all else' and easily recognized the desirability of having an open, transparent relationship with valued customers. For this person, predisposed to be trustful, the risks were outweighed by the gains and his firm had regained lost market share through its partnerships with specific Japanese customers. His view of fellow German supplier companies was they would have to learn to be more trustful in the new era of globalization and increasingly internationalized supply-chain relationships characteristic of the 'lean production' approach to management. It can be argued that the higher the level of trust, the lower the transaction costs associated with contractual, legal or insurance-related safeguards against relationship breakdown due to opportunistic behaviour by one of the partners. Fear of the loss of crucial know-how, without the preservation of which the firm will lose its competitive edge, is the main reason for reluctance to be completely transparent. Hence, the trustful supplier may have had little to lose through trustful transparency, but much to gain, while the reverse was the case with the Singaporean firm. Alternatively, in an era of rapid innovation, the sceptical supplier may have been over-cautious. The increasing availability of information capable of being translated into usable knowledge and the high degree of inter-action among networks of innovators may make being a distrusting entrepreneur an economic handicap. Hence, the importance of learning to trust.

Sako's (1992) study of trust is the most relevant and thoroughgoing analysis of the kinds of issues raised in the foregoing discussion since her book is precisely focused upon trust in supply chains. Of her three key conceptual cate-gories of trust, the first, Contractual Trust seems, at least partly, to capture what is going on. Contractual Trust is said to involve keeping oral or written promises, and if oral rather than written promises are kept, that is a sign of the highest kind of Contractual Trust, though not necessarily the strongest kind of trust, which presumably requires no contract. The above examples are not instances of Sako's other two categories, the first of which is Competence Trust, based on the belief that a supplier can do the job competently and reliably. This, it has been suggested, is not necessarily a category of trust at all but more in line with Luhmann's (1979) definition of 'confidence' which can be distin-guished from trust. Confidence is, in important ways, a matter of technical judgement that a firm or a person can undertake to fulfil a task because they have the skills, equipment and organizational capability to do it. Trust is more akin, as Sabel (1992) suggests to not taking advantage of the other's vulnera-bility. A person expressing confidence in someone else's competence may not need to supply information or knowledge that allows them to realize their vulnerability. Indeed, vulnerability may not enter the equation since a confi-dence relationship need not rule out a second or third such link as a means of insurance. This is probably true of a trust relationship in that more than one can easily be envisaged, but to share the core knowledge of vulnerability with too many could be debilitating, and even bring about the realization of that vulner-ability. Oddly, *confidence* may not require *confidentiality*, whereas trust nearly always implies some degree of it.

Sako's third category is Goodwill Trust, where there are mutual expectations of commitment to the relationship, where there is substantial give and take on both sides, but which, she argues, requires both of the other two types of trust to exist already if it is to be sustained. Goodwill Trust is thus what follows if the books have been opened, the firms like each other's culture and competence, they become mutually dependent and carry on doing business without contracts or with minimal contracts. Thus, in instances where, in other cases studied in the Cooke *et al.* (1993) report, suppliers had opened the books to a key, usually Japanese, customer, there would typically be a period of time when the supplier was having to adjust to the new quality or reliability demands of the customer. The latter would lend technical or other personnel or equipment to the former to help them learn how to meet the required standards. In a long-established preferred supplier relationship this would become relatively common and might reverse, such that the customer was learning from the supplier's innovations at prototype stage. Interactive innovation of this kind tends to be the kind which benefits substantially from 'economies of proximity'. More will be said about this elsewhere in this book but for the moment we need mainly to consider the advantages to customer and supplier of being able swiftly to organize emergency meetings between engineers, technicians, marketing personnel, etc. to discuss and jointly resolve problems or opportunities arising from the development or even the design of prototypes. If the partners are co-located, the reductions in uncertainty, time lag and transaction costs are clearly palpable. Goodwill Trust is thus a creative source of 'learning capital' which is capable of being constantly invested in the operations of those who are partners in the relationship.

But when does Goodwill Trust, or indeed any of the other kinds of trust for that matter, become a problem rather than a solution? This too will be discussed further when we tackle the relationship between 'economies of proximity' and 'learning capital' but suffice it to say here that there is an interesting literature inspired by the work of Granovetter (1985) and followed up by Grabher (1993) which speaks of the dangers of introversion in, for example, innovation linkages between firms. Thus, while Swiss watchmakers were interacting to produce greater reliability from ever-more finely chiselled mechanical technology, Japanese firms had solved the time-keeping problem through the use of quartz-based electronics. In turn, the new technology decimated the Swiss industry because of its massively lower cost. The Swiss had become 'locked in' to jewels and springs because of what Granovetter calls the 'weakness of strong ties'. When technology and associated economic development are in a reasonable degree of synchrony, and the institutional supports such as skills provision from schools and colleges, investment capital from bankers who understand mechanical watch technology, but would fear microelectronics, and technology transfer agencies have learned how to relate to reasonably predictable industry requirements, everything runs smoothly, like a well-calibrated Swiss watch, as it were. But, if some 'left-field' type of innovation is projected into the *milieu* (as Maillat, 1995, calls it) then the system may lack the 'absorptive capacity' (Cohen and Levinthal, 1990) to be able rapidly to learn the new lessons and respond practically. Social economists like Granovetter and Grabher advocate relationships which may realize 'the strength of weak ties' through linkages based on 'systemic loose-coupling' to avoid lock-in. What does this mean? Primarily it means developing many kinds of high and lower trust relationships both locally and globally. This does not mean high trust local and lower trust global linkages, it involves high and lower trust relationships at both levels, and those in-between.

This implies a high propensity for experimentation, on the one hand, and monitoring, information-processing and knowledge extraction on the other. Networks, which will also be a focus elsewhere in this book, require network management of the kind described in Cooke and Morgan (1998). Is this something that requires special skills and a specific appointment, or division, in a firm? Probably, until the majority of employees who need to network have learnt the rudiments of managing the information that inevitably floods in under a 'strength of weak ties' regime. Thus, experiments in the Netherlands, Germany and parts of the UK like Wales, with the idea of 'innovation assistants' being recruited from the recently graduated cohort of university students to enable small and medium-sized enterprises (SMEs) to handle new learning demands occasioned by globalizing markets, supply chains and information highways, have proven successful. This, particularly in contexts where existing management has no time to learn network management and judicious government agencies can also see a way, with modest subsidy, of reducing graduate unemployment rolls, is a sensible way of dealing with the exigencies of the learning economy. But in the long term, such functionaries will become a new elite if the skills they bring from modern higher education cannot, at least in part, be shared with their colleagues in a trustful, learning environment.

Trust, learning and knowledge

The problems that can arise in an economic context such as a firm, in which truly innovative knowledge gives rise to a situation in which management has absolutely no understanding of the technical reasons why particular practices work successfully, can be debilitating. Management that, perforce, trusted staff who either understood the knowledge, or understood enough to mimic those with the knowledge, while operating on a globally networked scale, star in the account of the demise of Barings, the UK merchant bank, provided by Gapper and Denton (1997). In one sense, the key to this story of greed and incompetence is represented in the award of the 1997 Nobel Prize for Economics to two American mathematicians, Fischer Black and Myron Scholes, who, in 1973 invented an equation which enabled the value of options to be determined with consistency. Oddly, these authors make no mention of Harvard professor Robert Merton (son of the famous sociologist) who solved the equations, using hydrological turbulence theory, later known as chaos theory, that made modern futures trading possible. He also won the Nobel Prize, with Scholes (Black died before receiving the award) and both established, with John Meriwether and others, the notorious 'hedge fund' company Long-Term Capital Management (Lowenstein, 2000).

Options are a type of 'derivative' or financial contract which derived their value from something else, such as the future value of something. Classically, 'pork bellies' would have a tradeable future value on the Chicago stock market, where they were invented to enable farmers to protect themselves against fluctuations in demand. A farmer could fix a price in advance of producing a commodity (e.g. wheat) by making a deal (or a gamble) on the price at a future, agreed date. If the real price was lower, the farmer was protected, if it was higher, the trader's gamble had paid off, but the farmer still realized his agreed return. Futures are ways of hedging bets, options offered the farmer the choice of whether to sell the commodity at the agreed price, or sell it at the market price, something he would clearly prefer to do if the market price was higher than expected. Of course, for the privilege of the option, the farmer had to pay a premium to the trader. If the premium cost more than the return from exercising the option, the buyer of the option lost money and vice versa. Buying and selling options became a market in its own right, detached from the commodity producers themselves as, early on, farmers realized they could trade the options separately from the community. Merton, Black and Scholes solved the problem of how to know the right amount to charge for the premium to offset the risk of the option price being lower than the market price.

Traders who understood the Black–Scholes and related models were recruited first by American banks from university mathematics and engineering departments. They were known as 'rocket scientists' who understood how to calculate derivatives prices and engage successfully in an apparently risk-free practice of arbitraging, meaning exploiting inefficiencies in markets between futures and actual prices. Initially, few banks understood how arbitraging worked, and within those that did, only a few people, namely, the rocket scientists, understood how to do it. By the mid-1980s financial futures markets had evolved in which trading in shares and bonds replaced commodities trading. The key growth market of the period was South-East Asia, particularly Japan, and Barings, like other banks, traded derivatives on Japanese markets. Barings' chief rocket scientist in Tokyo concentrated on buying Nikkei shares and selling futures and vice versa. The best prices for the latter were found on the Singapore International Monetary Exchange. A junior trader there was Nick Leeson whose skills lay in buying futures quickly at good prices, both of which enhanced profitability for Barings.

Leeson was not a rocket scientist but a trader and former back office worker who was employed to get good prices and settle trades by ensuring the bank received all its purchased financial contracts. Baring Securities had established a small operation called Baring Futures (Singapore) and Leeson was selected to manage the office. Because it was a small office and, presumably, 'competence trust' or confidence had been placed in Leeson, he was, fatally, given the two jobs of trading manager and settlements manager. As Gapper and Denton put it:

> Banks usually separated traders from the back office to prevent fraud. Traders were often tempted to cheat or cover up mistakes, and this was usually caught by the back office. But it had not seemed worthwhile to have both a head trader and somebody in charge of the back office at Baring Futures.
>
> (1997, p. 12)

Leeson had quickly made his reputation as a good back office settlements clerk but he was now engaged in buying and selling futures too. Taking orders to buy and sell from Barings Osaka and Tokyo traders Leeson had noticed that big orders had the effect of moving prices under their own power on the smaller Singapore market. This, in turn, created an opportunity, for, by using the bank's own money, he could buy futures in Osaka and sell them to the customer at the Singapore price before it moved. Customers preferred to use the Singapore exchange because it was cheaper than the Japanese exchanges. The bank received its profit risk-free from the difference between the lower price paid for futures in Osaka and the higher price for which they were sold in Singapore. This 'switching', as it was known, earned more money for Barings in 1994 than did the rocket scientists. Competence Trust in Leeson's abilities thus grew, but did not occasion investigation because it was not unusual for someone with 'the golden touch' to emerge among the bank's traders.

Trading successfully on a very large scale required only that there existed a source of cash to trade with. Keeping daily track of banks' debts to the exchanges, or 'margining' was what the back office did. But because of the extent of his trading activities, he was starved of the time needed to do this properly, thus undermining the Contractual Trust between bank and trader, unbeknown to the bank. This casts an interesting, critical light on the view that contractual and conceptual trust reinforce each other to give rise to Goodwill Trust. Here is an instance of Competence Trust leading to a break-down in Contractual Trust because of an over-reliance and, as it transpired, over-estimation of Competence Trust. It could be argued that Goodwill Trust was what characterized the culture in merchant banking, developed, at least in Barings, over generations, and that, on the basis of results (however mysteri-ously achieved), Competence and Contractual Trust followed. Thus, we have Sako's analysis of the requisite priorities between types of trust to achieve successful outcomes placed in reverse in this case, something which tends to vindicate her analysis by counterfactual means. Clearly, the question of compe-tence *per se* projects itself in relation to the organizational management style of the bank in question. From the point when it '... had not seemed worthwhile ...' to separate the settlement and trading functions, to the assumption that the massive profits (£28 million) emanating from the activities of one lowly trader were risk-free, because the bets were presumed hedged, the bank showed staggering incompetence. Once, a debt of £50 million had apparently failed to be collected but was explained away to the auditors, something which led to Leeson being required to reduce his switching activities. The bank sent an investment executive to find out how these profits, but also the large cash-calls, were generated, but 'It was the end of a tiring day and Norris was tired. Leeson was sweating despite the air-conditioning, and he launched into a long, rambling explanation of his business, full of jargon that Norris did not under-stand' (Gapper and Denton, 1997, pp. 15–16). Norris (the investment executive) would receive a 1994 bonus of £1 million and the only doubt he expressed was about Leeson's being paid £450,000 in case it 'spoilt' him.

Nevertheless, the £50 million shortfall in collected debts had triggered an investigation process. It should be recalled that the only cash-flow *monitoring* was the daily accounting, sometimes incomplete, of outstanding margin calls. Closer investigation showed the Barings' records of Singapore trading to be extremely poor, and the shortfall to be £95 million. The Futures and Options

Settlement Clerk sent from London had failed to get a clear understanding of what was going on when a new margin call for $45 million reached London indicating Leeson had not reduced his switching. Even the Treasurer of the Baring Investment Bank was unable to understand why so much cash was required for margin calls; numerous other responsible executives had been called to Singapore:

> As [Norris] looked around the room he could see nearly all those responsible for Leeson. It struck him this would be the first time they had actually seen the Simex (Singapore Futures Exchange) statements themselves. Everyone had relied on Leeson to tell them what he was doing. But he was no longer there, and they were in the dark without him.
>
> (ibid., pp. 20–1)

In an error account (the infamous Account 88888) were found Nikkei futures worth £3 billion all dependent on the Japanese share index rising above 18,500 to yield a profit. But after the Kobe earthquake it had fallen to below 18,000 and was not expected to rise. As a consequence, Barings owed £200 million and the futures or bets were not in fact hedged. Leeson had been making huge margin calls to buy more and more Japanese shares and move the market. As his futures transactions had failed to do this, he had started trading options for the bank, something he was not allowed to do. The damage from these activities was a £100 million loss to Barings which, added to the £250 million from the futures *débâcle* entirely accounted for the balance sheet value of the bank; accordingly, the bank was now bankrupt. An attempt was made to bail out the bank by drawing on the Goodwill Trust of the merchant banking community in London, but on finding that margin calls had reached £700 million:

> By this time, most of the bankers' previous goodwill towards Barings had evaporated. They were being asked to risk their shareholders' money to rescue a bank which had paid inflated bonuses to directors who were so ignorant of what was going on under their noses that they handed £700 million over to a crooked trader to help him defraud them.
>
> (ibid., p. 51)

Oddly, Leeson's fraud was not embezzlement but an attempt to hide failure to earn continuing profits for his employer. As a case study in the failure of corporate culture, Leeson's practices contributed the apotheosis of closure, secretiveness and ignorance. Yet strangely, the bank's fortunes were based on a recklessly high level of Goodwill Trust. As noted, this inheritance from reputational judgement and customary practice predominated over more practical Contractual and Competence forms of trust, themselves subject to little or no test other than results, measured in terms of the single performance indicator of profit.

This kind of trust can clearly leave a company exceedingly vulnerable to self-devouring transaction costs where decision rules, communicative openness, a learning propensity and systematic monitoring and feedback arrangements are absent. Goodwill Trust may be significantly transaction cost-reducing where it is undergirded by Contractual and Competence Trust but it is clearly no substitute for the fulfilment of trust of these other two kinds. This is a useful corrective to the view that may be in danger of evolving to the effect that trust is a kind of alternative economic rationality that might even replace more orthodox rational calculation because it is, apparently, so much more economically rational owing to its transaction cost-saving capacity. Trust occurs between human beings and, even if not for reasons of venality, humans make mistakes, which, especially in a cut-throat, competitive and individualistic environment, they may well seek first to cover up rather than disclose. Trust, therefore, may envelop an apparently *disembedded* environment, one which can appear like a war of all against all and the opposite of the *embedded* setting alluded to in Chapter 2 and to which we return below, but despite enveloping it need not pervade, systemically, the interior of an organization in an apparently trustful environment where goodwill is often seen as a priceless asset. As we have seen, goodwill can have a price that it is painful even contemplating the risk of having to pay. The Barings case gives an indication of the price the wrong kind of trust can entail.

Distance creates problems, particularly for traditional Goodwill Trust of the kind characteristic of Barings Bank, in the absence of decision rules, or their adequate enforcement, adequate decision points where information is assimilated and knowledge learned and acted upon, and the monitoring and feedback processes which supply information. The paradox is that all of these arrangements appear as the antithesis of trust; they are mechanisms put in place where the presumption is made that trust will be breached. There are expensive transaction costs in setting up control systems of this kind, which trust is meant to circumvent. As we have seen, the 'breach' of trust that occurred in the Leeson case was competence-based. The actor in question possessed neither the levels of knowledge needed to perform the functions actually being executed, nor the internal controls, acquired through training in the case of the rocket scientists, to recognize the dangers of what he was doing. Nor did the bank have the controls to monitor and act on this situation at the time it was developing towards catastrophe. And this was all going on a long way from head office.

Trust is intimately connected with communication. Good trust relationships lead to improved communication, not so much in the sense of quantity, but effectiveness (Deal and Kennedy, 1982). To revert to Sako's (1992) categorization of trust: where only Contractual Trust is present, expectations are that actors will, minimally, adhere to formal written rules, meaning precision of communication is vital. Precise communication can take the form of 'codified knowledge' whereas the looser form of 'tacit knowledge' to use Michael Polanyi's useful distinction (Polanyi, 1966; see also Cooke and Morgan, 1998) is more common in a Goodwill Trust context. Distance does not degrade the message significantly where it takes the form of codified knowledge but obviously must where change is rapid and technical ignorance is high, in other words, where whatever rules exist are inadequate or easy to breach. In such contexts Competence Trust is at a premium since the combination of discipline from training and creativity, innovation and judgement from social learning and intrinsic capability are the only day-to-day controls available. It could be concluded that the problem of distance compounded the implicit weaknesses of a Goodwill Trust organization making an essentially Contractual Trust agreement (more normal to arm's-length exchange relationships) with an actor likely, through insufficient training or capability, to prove incapable of maintaining Competence Trust. Failures of communication from both organization and actor, desperate breach of the, weakly enforced, decision rules of the organization and reliance on inadequate tacit knowledge were merely the symptoms rather than the causes of such a major malfunction as the Barings' bankruptcy.

Learning economies, whether they are particular firms, the people they employ and the other firms and organizations with which they interact, or the wider community of economic actors focused on industrial sectors or even 'clusters' at the regional, natural or global levels, are those that recognize how reliant upon trust they truly are, but understand that to over-regulate relationships with contracts and rules is both expensive and trust-destroying. Consequently, they seek to create a climate of openness, communication and reflexivity (e.g. 'benchmarking') within a broad framework of consensually derived rules and 'acceptable practices' or conventions that are widely understood. In this way an 'obligational' culture may be stimulated in which trust rather than formal rules and required actions motivate action. To repeat, this is harder to achieve outside the confines of the firm than inside it (one of the reasons why firms exist, according to Coase, 1937) and at a distance from rather than in proximity to other key intra- or extra-firm or organizational actors. Learning is the central process promoting openness, communication, trust and the shape of decision rules where it is inclusive, accessible and based on reliable knowledge, but it must be both locally and globally attuned.

There is one further paradox to resolve at this stage of the discussion of the relationship between learning and trust. Earlier it was noted that trust always implies some degree of confidentiality whereas later it was concluded that a trustful organization, like a high-trust economy or society required openness, communication and inclusivity. The point here is that not everyone wants or needs, or indeed is capable of, knowing everything that goes on that may, in unpredictable ways, be of relevance or consequence to them. This is, clearly, a particularly sensitive issue with respect to proprietory knowledge. Firms and individuals within them have an understandable antipathy to sharing expertise that can be appropriated by a competitor. Typically such knowledge is protected by confidentiality agreements, where sharing it with a competitor would be deemed likely to render the firm vulnerable. Intellectual property rights, patenting law, copyright, and so on are further institutional protections for expertise. The openness of a trustful organization or economy, though, is more procedural than substantive. An interesting example of the provision of sufficient information about arcane processes arises with 'citizen's juries', where ordinary citizens are 'trained' up to a level sufficient to enable them to make intelligent judgements about policy issues. 'Deliberative polling' is a similar technique used in the academic research context. This is the degree of openness that is required to help firms or organizations overcome the problem of top management being wholly ignorant about the means by which they are successful and might, suddenly, become the opposite.

Trust, social capital and development

It has been argued, thus far, that learning economies will tend to be those in which trust among economic and social actors is pronounced and that trust and learning can have self-reinforcing effects in terms of improved use of resources, both in relating to economic efficiency and effectiveness. These flow from reductions in transaction costs and increases in information economies and associated knowledge-effects. It has also been shown that trust is a dangerous medium where it is insufficiently studied, analysed and deployed. 'Opening the books' is acceptable to some, not to others and operating at the leading edge on the basis of Goodwill and Contractual Trust without adequate Competence Trust can be disastrous. Yet the more business interactions involve supply-chain relationships, or more simply, sub-contracting or the use of contingent labour, the more care has to be taken of the factor of trust at the interface between the legal entities who are party to any agreement. In principle, high-trust societies should benefit economically from not needing to engage in transaction costs to insure against breaches of trust. But where are such societies, and if they exist, how are they formed? More to the point, can low-trust societies learn their attributes and translate them into economic success for themselves? These questions have exercised a number of authors recently and, as Humphrey and Schmitz (1998) note, the World Bank established a study group on this theme in 1996 with a view to tracking down a possible missing ingredient in economic development.

The last-named authors argue that trust, in the context in which it is being discussed here, is closely intertwined with *sanctions* and that those countries that have clear and strong systems of sanctions, for example, against late payment of debts in Germany, also have strong trust relationships in the economy. Intermediaries, such as business associations, play a secondary but important role in facilitating the flow of business information and even standards among sectoral communities. Together, these create conditions for 'extended trust', necessary to the management of complex, externalized inter-firm relations of the kind under investigation here. These things are clearly important, but counterfactuals exist, notably Italy where the regulatory and formal associational environments for industry are probably less robust than in Germany but extended trust, in the small–medium business sector especially, can be extremely high. In the Italian case, *sanctions*, to the extent they are specifiable, are probably social, or community-centred, as much as economic, something to which Humphrey and Schmitz do, in fact, allude elsewhere in their contribution in respect of 'industrial districts'.

Mention of Italy and the question of trust, social capital and development takes us onto the research of a number of scholars, one of whom, Fukuyama (1995) places considerable emphasis on the Italian experience in his wide-ranging, but also somewhat confused and confusing, analysis of trust. He seeks to answer the question posed to Ricardo by Malthus, as to why some nations become rich while others remain poor, by reference to differential levels of trust within cultures. Low-trust cultures have low social capital and, while such cultures may develop economically, firms tend to remain small and globally uncompetitive unless supported by their state. Italy is an acid test for this thesis, which it has to be said, scarcely impresses by its robustness. Fukuyama wants us to conceive of Italian culture, like Chinese, which he also presents as low-trust, as over-focused on the binding ties of the family and inadequately formed in terms of its development of *associationalism* within civil society (on association-alism, see, for example, Hirst, 1994; Cooke and Morgan, 1998). This he contrasts with American, German and Japanese culture which are said to be strongly associational, in the sense that many members of society are active members of clubs, groups, citizen's leagues and the like, or as businesses are members of trade associations, high in trust, social capital and economic perfor-mance indicators.

Fukuyama's problem, which is not adequately tackled, is that the 'Italy' he appears to have in mind is the South. He draws heavily, using lengthy quota-tions, from Banfield's (1958) study of Montegrano in Sicily where 'amoral familism', or the culture of moral obligation to the family alone, predomi-nated. Thus:

> The small family firms of central Italy nevertheless constitute something of an anomaly ... Why do small family firms predominate in central Italy ... ? The high degree of social trust in this region should have allowed producers to go well beyond the family in business organizations.
>
> (Fukuyama, 1995, pp. 104–5)

What this reveals is a linear mode of thinking whereby trust creates social capital, which promotes development, which is represented in large, preferably multinational firms. This is a particularly American obsession, shared by both self-confessed neoconservatives like Fukuyama and more left-of-centre authors such as Bennett Harrison (1994) who was sceptical that the Marshallian clusters of north-central Italy could survive the tendency towards acquisition and merger so prevalent in the American economy.

Fukuyama begins to glimpse an explanation for the 'anomaly' of north-central Italy's successful small-firm industrial districts when he refers to *networks* of firms but in the process loses another major plank in his argument by noting that these are not based on *familism* but on a social and technical division of labour. So, central Italy has few large firms, low familism of the Sicilian kind, yet high trust, high associationalism, high social capital and very high economic performance indicators. Two elements that are missed entirely are the regional variations within Italy, such that to try to read Italy today from an account of Sicily in the 1950s, is to pile error upon error, and the politics of small-firm industry as between regions. The 'red belt' of Emilia-Romana, Tuscany and Marche even have different strands of left politics and policies between them, the first decentralist with enterprise support, the second centralist without enterprise support and the third somewhere in-between. Elsewhere, in Veneto, Lombardy and even Fruili, politics are conservative, policies are close to *laissez-faire* but local business associations are powerful and influential with politicians, both regionally and nationally. The style of small-firm networking varies regionally, with Benetton dominating extensive supply chains in Veneto, but Emilian clothing firms forming less well-known groups of small enterprises that cover a range of market niches but supply well-known American and European department stores with low-cost, high-quality products. It should also be added that by the 1990s, some regions in the South, notably Apulia and even Basilicata, have begun to detach themselves from the economic stagnation that still afflicts Calabria and Sicily. There Fukuyama's and others', such as Humphrey and Schmitz's, point is still borne out that where civil society scarcely exists, except for the church, and the state has little or no legitimacy, criminality thrives and breeds, and only the nuclear family has affective meaning.

Perhaps Fukuyama's difficulties in explaining the economic success of much, if not all, of modern Italy lie also in his attachment to Max Weber's (1930) *The Protestant Ethic and the Spirit of Capitalism*. This book seems to be experiencing a revival given that Landes (1998) also draws heavily upon its insights in his attempt to answer Malthus' question. As Martin Daunton (1998) points out in his review of Landes:

> The message of the book is that 'what counts is work, thrift, honesty, patience, tenacity'. What is needed is to live for work, an experience of a 'small and fortunate elite' that can be joined by anyone willing to 'accentuate the positive' even when they are wrong, in order to learn from mistakes and improve in the future.
>
> (1998, p. 24)

Non-Protestant religions hampered the development of intellectual freedom and curiosity, markets, secure property rights and political freedom. Those countries that had religions, and cultures related to them, that did away with such barriers, including Japan (after the 1868 Meiji Restoration) or overseas Chinese communities, have prospered. As Daunton concludes:

> [This] comes nowhere close to explaining the wealth and poverty of nations by their culture, for the argument of the book is no more than tautology. Japan was economically successful in the 20th century. How is this success explained? By its culture. How do we know that culture was important? Because Japan was economically successful.
>
> (ibid.)

Clearly, as with Fukuyama's approach, this explanation of economic development has a number of problems. First, to repeat, some non-Protestant economies have grown rapidly, notably Japan, Italy and, recently, Ireland which forces a displacement from Protestantism to the 'work ethic' of individuals who 'live to work'. Second, some Protestant countries and communities, having shown inventiveness and growth capability have revealed how difficult it is to sustain those characteristics. Both the UK at the macro-scale and communities like the Shakers in the USA have declined relatively or absolutely in these terms despite having produced goods in their heyday which are still in high demand. And third, neither pays much attention to contextual factors as stimuli to growth, such as the devastation following war and the injections of foreign aid, or simply capability to produce and sell goods cheaply on world markets because of having to start from 'ground zero' to survive.

Before examining some contemporary examples of the problem of development from virtual ground zero, it is instructive to examine the recent success of another non-Protestant country, the Republic of Ireland, that, on the face of it, meets none of Fukuyama's success criteria. First, it is an overwhelmingly Catholic country with over 90 per cent of the population confessing that faith, in the economic sphere it has not been marked out in studies as being particularly trustful, rather, being the opposite, individualistic, combative, and in the small firm sphere not especially globally competitive. Social capital could not, as a consequence, be said to be especially high, though in the socio-cultural sphere, there are many Celtic associations for music, dance, literature and sport, often quite closely linked to the Church in terms of organization. Nevertheless, Ireland has displayed the highest Gross Domestic Product growth rate in the 1987–97 period of all the advanced economies in the OECD, touching 10 per cent per annum on occasion and averaging some 6–7 per cent per year. Ireland also has some rather large indigenous and foreign companies.

How is this to be understood? Leading economists such as Kenneth Arrow and Paul Krugman, in contributions to Gray (1997), identify, *inter alia*, macroeconomic stabilization through trustful social partnership agreements on wages, investment and inflation targets between capital, labour and the state, judicious use of European Union financial assistance to build up human capital, especially in information technology and software skills, and a favourable corporate tax regime aimed at attracting foreign direct investment capable of making use of the skills available. Uncertainty, indebtedness and weak human capital (including entrepreneurship) were diminished in a strongly state-led manner and under such relatively propitious conditions the 'Celtic Tiger' flourished, meaning investment opportunities were perceived and taken up relatively swiftly. Interestingly, it is also *post facto* that trust-building and social capital initiatives are being taken, through state-led experiments in promoting collaborative practices through supply chains or horizontal inter-firm networking programmes (see, for example, NESC, 1998 and Cooke, 1996) aimed at helping to sustain economic growth and business performance for the future.

Further research to ascertain the likely receptivity of Irish industry to a policy of promoting Porter's (1990) notion of industry clusters (Clancy *et al.*, 1998) found that in three industries studied that were internationally competitive (dairying, software and music) a Porter-style strategy would be inappropriate. This was because the Irish case showed a number of divergences from Porter's suggested determinants for competitive advantage. Ireland's small scale means that there is limited domestic demand, domestic rivalry, domestic supply and a significant role for foreign direct investment, something Porter explicitly excludes from his schemes. Accordingly, the industries examined were found not to exhibit strong cluster-like characteristics. However, this is more by way of a test of Porter in the Irish context than a test of the extent to which there is interfacing between firms in complementary industries or activities. When this was explored, it was found that there are appreciable connections between firms and that benefits could accrue from promoting more. This was clearest in the indigenously owned part of the software industry centred in Dublin where there exists a high degree of growth-enhancing interaction through sub-contracting, joint product development, technical exchange, and the like. Even in the largely indigenously-owned dairy industry, information flow and knowledge transfer among industry actors were occurring and perceived by the industry to be valuable. One conclusion of this work was that such informal networking should be strengthened, but also, more intellectually, the restrictiveness of the Porter model of successful clustering should be questioned as an economic development strategy.

In a review of the findings of Clancy *et al.* (1998), O'Donnell (1998) offered alternative reasons for the relative absence of Porterian conditions from Ireland's economic success. One alternative could be that the evidence of informal networking was a sign of 'nascent clustering', thus it was too soon to be confident that clusters would develop, but logically Ireland's small size should still make it unlikely in Porterian terms. The other alternative was that Porter is wrong and that intense domestic rivalry and a strong 'home base' are not necessary to the competitive advantage of nations. That is, Irish companies may be benefiting not from external economies derived from business interaction within Ireland but through transnational linkages. In other words, Ireland's experience may herald a new, post-Porterian model of economic development with new possibilities for newcomer economies or late industrializers. The point here would be that the Irish economy is so open that Porter's emphasis on the localized nature of the skills-creation and innovation processes may not apply since external economies such as these can easily cross national boundaries because of foreign direct investment.

A comparable argument, though not one in which foreign investment plays such a part, is made for the small-country economy of Denmark by Lundvall and Johnson (1994) when they discuss the necessity for such small, open economies to be 'learning economies' by being well attuned to developments in respect of innovation skills and business organization occurring outside the country and using such knowledge to learn competitive advantage. This is necessary because small economies cannot possibly cover all the sectors or angles in, for example, public promotion of research and development. Hence, the imperative of inculcating a learning propensity in the governance and business organizations of the country and an open, networking mentality among small firms to ensure such knowledge flows systemically through the economy is appropriate. Clearly, such a recipe makes eminent sense for regions within larger economies that lag behind the core, often capital-city region, of their country.

So, it is evident that trust and social capital, not to mention inter-firm and firm–agency learning of the kind said to be paradigmatically the case in areas such as north-central Italy, are less than wholly persuasive explanations for the recent success of an economy, Ireland, not imbued with the Protestant ethic. However, further examination suggests that trust may, nevertheless be prefiguratively, nascently or latently present but not in the manner that Michael Porter, one of the leading proponents of clustering for competitive advantage could have predicted. This is important, for in a fairly obvious sense, relatively few regions or agglomerations or cities in the world can be globally competitive in a given industry based on a strong domestic market and substantial inter-firm rivalry since most do not have the requisite scale. The Irish exception may not be alone inasmuch as comparable processes (though not at the same levels of GDP growth) appear to have occurred in regional economies such as those of Scotland and Wales in the UK and, perhaps, Catalonia in Spain as well as, to a considerable degree, the Pacific Rim economies that began to take off following inward investment decisions from American, and even more so, Japanese firms in the 1970s and 1980s. In all cases a substantial presence of inward investment has bolstered these economies and real opportunities exist for domestic trading by host-country firms of the requisite quality and expertise.

But what is it about trust and social capital in the cultural sphere that may make it valuable in the economic sphere? We noted in the Irish case that there is a rich associationalism in the cultural sphere but we simply do not know whether or how that might carry over into the economic sphere. Of course, it is not unknown for business people to conduct negotiations and make deals on the golf course but we are posing the question more structurally. Is there something about a society which has many people active in a variety of associations that 'trains' them to make use of that experience in the economic sphere even when they might be interacting with wholly different individuals in the two spheres? One author who would answer that question in the affirmative is Robert Putnam (1993) who explored the role of civic institutions in economic development in what, once again, constitutes almost a contemporary laboratory for the study of civic associationism, modern Italy.

Putnam's task is by now well known. It was to try to understand the Malthusian question of what makes rich and poor places by reference to possible variations in civic attachment, or what he called (after Jacobs, 1961 and Coleman, 1990) 'social capital'. It is also well known that he found regions in northern Italy to be generally more civic, more associational, better organized administratively and more prosperous than those in the south. This was even though, in some cases, those same northern regions had been less developed in terms of health and economic indicators than some in the south at the beginning of the twentieth century. His conclusion was, in brief, that social capital was responsible, through civic associationalism, for good regional economic performance *rather than* that good economies gave rise to good social capital.

However, the question we wish to explore here is how does the social translate into the economic performance? In truth, this issue is not a strong point of Putnam's analysis. Civic involvement includes only one 'associational' variable out of five and that measures longevity of associations rather than type and membership. Socio-economic development is taken as shares of agricultural and industrial employment. Putnam finds high positive correlations between civic engagement and socio-economic development, but with a seventy-year time-lag, at the regional level. Of course, other contextual variables are highly likely to be involved in changes to both civic engagement and economic development. And, in response to his own question: 'Through what mechanisms might the norms and networks of the civic community contribute to economic prosperity?', Putnam concludes more research is needed. When he does begin to explore possible relationships he draws particularly on central Italian regions like Emilia-Romagna where there is both associationalism in the civic and economic arenas. In this region though, both kinds of association are affiliated to political parties like the old Italian Communist Party (PCI) or Italian Socialist Party (PSI). The promotion of a political culture of decentralist socialism, projected through a vibrant, collective cultural life and a communal approach to SME development and support was, and remains part of a political project. A different, although still PCI project, operated in Tuscany. There, the dominant party was much more centralist and saw SMEs as anachronistic. Thus, entrepreneurial associations of a private nature were formed to facilitate the functioning of industrial districts, against the hostility of the ruling party at regional level. Hence, it is misleading to distance associationalism from politics too much in the Italian case, and maybe elsewhere.

It is conceivable that cultural and economic associations in particular regions have overlapping membership but it is probably unreasonable to expect this to be extensive. Entrepreneurs regularly state they are short of time as well as funding in surveys of small business practices. A recent survey of business networking initiatives promoted by UK Training and Enterprise Councils found this to be the case and, further, found that of the approximately 40 per cent of firms that met Granovetter's (1973) definition of 'weak-ties' relationships as meeting more than once a year but less than twice a week, only 15 per cent interacted as such in social or leisure-related settings, and none of those in 'strong-ties' relationships met in such settings. Much of the most important settings for the former are professionally focused 'organized workshops' and 'forums or conferences' while for the latter it was just forums or conferences (Huggins, 1998). Moreover, for both, actual one-to-one business meetings were more important forms of business-to-business contact than either of the other, more extensive, forms. This should not surprise us, associational activity is a means of enabling information to flow and, potentially, through collective action, influencing powerful organizations such as governments to support activities, perhaps financially, that further associational interests. Experiences gained, or heard about in getting the municipality and perhaps others to support good cultural cases are simply 'in the air' to use Marshall's phrase and may give confidence to try out the same approach in the enterprise support sphere, or, of course, vice versa.

Finally, this question of the functions of social capital in forging trust relations through associationalism has importance for policy thinking and action in pursuit of economic development. In this discussion we will look at recent changes occurring in Central and Eastern Europe, economies that have shifted dramatically from centralized state planning to liberal market co-ordination without in either case developing the density of intermediate associations between state and economy that have been under discussion thus far. In the process, we will reconsider the dimension of innovation with which this chapter began. It is frequently argued that one key reason why the Soviet system collapsed was because of the poor levels of either social or technological and other economic innovation. While there may be general truth in this view, with respect to specifics like military and space technology, for example, the Soviet Union showed greater systems capability than all countries other than the USA, though cynics would say much of that innovation was copied from the USA and elsewhere. Freeman (1995) makes an interesting comparison between the national innovation systems of the USSR and Japan. He points out that, in some respects they resembled each other in having good, science-based education systems and the capacity to plan scientific and technological development. The USSR devoted more of its GDP in the 1970s to research and development (R&D), but 70 per cent of this was devoted to military or space research. The key differences were that the Japanese had strong integration in knowledge transfer linking R&D, production and imported technology while the USSR did not, and that the Japanese had strong user–producer linkages and sub-contracting networks, absent in the USSR. In other words, Japan was, as we know, the more networked and collaborative economy.

In their discussion of the probable importance of trust and social capital in economic development, Humphrey and Schmitz (1998) cite numerous authors and themselves state that: 'A major problem facing the transition economies is the scarcity, or absence, of the minimal trust required to facilitate transactions. We consider it the key obstacle to the establishment of an effective market economy in these countries' (ibid., p. 42). While it would be mistaken to reject this view completely, it will be argued that it is probably not universally the case throughout Central and Eastern Europe and may well be over-stated for some parts of the former Soviet Union if not for Russia itself. Having said that, it does seem reasonably clear that Russia is among the least well-performing transition economies and, among the popular arguments for that is the pervasive lack of trust there and the difficulty of reversing that condition. This deficiency is traced to inadequate cultural resources and social capital but also, as

Fukuyama and others note, ineffective law enforcement, meaningless regulations and rapidly changing economic policies. In such a context of massive uncertainty, ruthless profiteering and criminality are better rewarded than thrift and hard work *à la* Weber. Yet some of the old industrial elite still operate in new networks, and, as Humphrey and Schmitz show, deploy trust as a substitute for cash transactions in some periods of constrained cash flow. But they also quote an Italian author to the effect that the future of Russia may come to emulate the history of Sicily because of the absence of legitimized legal and property rights. The protection rackets rife in both settings are reminiscent of 1930s' Chicago, yet it should be remembered that Caponeism flourished in a modern country with well-established legal and property codes. Much of the gangsterism of the USA derived from a clash of codes – southern Italian against American in a context of high in-migration, personal mobility and ethnically rooted discrimination.

Research conducted by this author and colleagues into collaboration and partnership with respect to innovation among firms in Hungary and Poland of the former Soviet satellites and the Baltic countries of Estonia, Latvia and Lithuania as former component parts of the Soviet Union, improves understanding of the emergent business culture in hitherto low trust, low associational settings. In the work on Hungary and Poland, firms in key sectors such as automotive, electronics and mechanical engineering were asked to state whether they had introduced product and process innovations between 1993 and 1996, where the main sources of information relevant to these innovations lay and with whom they had engaged in partnership relations for the innovations in question and whether any partnership was formal (i.e. contractual) or informal (e.g. goodwill). In Hungary, an average of 46 per cent of firms had produced innovative products new to the market, small firms (employing less than 50) being the greater innovators (48 per cent), medium-sized firms (50–199 employees) the lesser (29 per cent). Only 22 per cent of Hungarian firms considered themselves process innovators, this time medium-sized firms (24 per cent) comprised the greater proportion of innovators, compared to small (22 per cent) and large (11 per cent). The key sources of innovation information were customers, journals, suppliers and conferences, in that order with consultancies at some distance behind. Main partners in product or process innovation were customers, suppliers, consultants and contract research organizations in that order, and formal contractual cooperations outnumbered informal ones by two to one (Makó *et al.*, 1997).

In Poland, 60 per cent of firms claimed product innovations new to the market, 40 per cent process innovations new to the market. Subsequent interviewing of representative firms revealed these innovations to be relatively modest, incremental or adaptive changes to existing products or processes or modifications of imported technologies hitherto not available in either of the two countries studied. Polish sources were journals, conferences or trade fairs, followed by customers and suppliers in that order; 70 per cent of responding firms depended on customers, 55 per cent on suppliers for innovation knowledge while the figures for journals and conferences were 89 per cent and 84 per cent, respectively. However, innovation partnerships were strongest with customers (67 per cent), followed by suppliers (55 per cent), then universities (33 per cent). Informal cooperations outweighed formal by approximately two to one. In both cases partnerships within the region where plants were located were stated to be the most important, followed by those elsewhere within the country (Galar and Kuklinski, 1997). Polish firms were thus more inclined to base innovation on codified knowledge, considered themselves innovative but were only modestly so, while Hungarian firms were more innovative and more reliant on non-codified or tacit knowledge. Both are respectable approaches to the Knowledge Economy, but social interaction is possibly the more likely source of novelty, especially if they become formalized as market transactions.

A comparable piece of research was conducted in the Baltic States, though the key industries subject to investigation were financial services and information technologies (actually mainly software and peripherals). In financial services (banks, insurance, business services, etc.) the most cited and highest-ranked formal interaction amongst firms was a 'preferred customer' relationship, though in Latvia 'outsourcing' was the most common. Informal relationships ranking highest in Latvia and Lithuania were 'networking' and 'informal knowledge exchange', with 'sharing facilities or personnel' second. In all countries these interactions were overwhelmingly centred upon the urban agglomerations of Tallinn, Riga and Vilnius. Some 30 per cent of financial services firms had generated product innovations new to the market in the period 1994 to 1997 and 34 per cent were process innovations, mainly organizational changes. Customers were the first source of innovation ideas, followed by conferences and consultants.

Baltic software firms, of which there are between 120 and 155 per country, mostly had their origins in the early days of illegal 'pirate programs' but regulations have been introduced to minimize this and now firms are mainly engaged in legitimate activities. Firms are small, few employing more than 50 persons, and growth was high between 1993 and 1997. Once more, a high level of business activity was focused within the capital cities – Riga (80 per cent), Tallinn (69 per cent), Vilnius (46 per cent). As with financial services top formal inter-firm interactions are with customers followed by suppliers and informal interactions are mainly 'informal knowledge exchange', 'informal networking' and 'sharing of facilities or personnel'. Relatively few innovations are new to the market but are modifications, adaptations and re-working of already existing Western programs. Customers, suppliers and journals are the main sources of information and innovation partnership. In each country, a software association had been established by 1997 and there was evidence that lobbying efforts by these had influenced governments to establish enterprise support and financial aid programmes for the software industry (Cooke *et al.*, 1998).

These experiences suggest rather strongly that, with respect to business-to-business activities, Contractual and Goodwill Trust relationships exist, perhaps in a rather narrow band of preferred customers and suppliers among firms of different kinds in transition economies. In the case of the Baltic States in particular, behind a competitive, arm's-length exchange form of market relationship are other, more cooperative, 'generalized reciprocity' (Putnam, 1993) relationships that are highly localized within the dominant urban communities. In turn, these have already given rise to the exercise of social capital in terms of the establishment of associative organizations, already lobbying governments with some success. Interestingly, in all cases, innovative activities appear to be highly compatible with inter-firm interaction, partnership and cooperation, particularly at urban or regional level, as predicted in the 'systems of innovation' literature, which stresses the peculiarly interactive nature of innovation related activities (see, for example, Edquist, 1997; Braczyk *et al.*, 1998).

Hence, by way of a reprise of this section, it can be safely concluded that trust, social capital and associational activity have their roles to play in the economic development process. Contextually speaking, it is undeniable that enforceable legal rules and legitimized property rights are prime requisites for business to flourish, though business activity is not necessarily made impossible by 'weak encoding' of these. While they are probably stronger in the Baltic States, Hungary and Poland, much of the interactive business activity occurring there has developed during years of equivalent uncertainty to that experienced in Russia. Criminality is present in these countries too, but, for diverse, partly historical, partly ethnic, mainly, perhaps, nationalistic and anti-Russian binding sentiment, economic activity seems less prey to criminality and nihilism than may be the case in Russia. Market-opportunity, especially towards the West, also gives an incentive structure that, to some extent, compensates for the virtual impossibility of raising investment capital from domestic banks due to prohibitive interest rates and capital starvation. Once business is embarked upon deploying skills from an accomplished education system, competition and co-operation become normal, everyday practices. Local social ties provide some degree of confidence that trust and obligation may be countenanced, and in this way social capital, perhaps as in Ireland or Italy, more strongly expressed traditionally in the cultural spheres of music societies, theatre and folk-dance associations, may come to be expressed also in business associations that link the economic dimension of civil society to the state. Clearly, there need not be a seventy-year time-lag between the former and the latter, although it is ironic that Ireland's miracle and the growth of some transition economies like the Baltic States are arriving about seventy years or so after their first political independences.

Conclusion

The argument developed in Chapter 2 of this book was that because of evolutionary processes, economies change in path-dependent ways, the original trajectory of which is not necessarily predictable *a priori*. Trajectories accrete around them norms, routine, habits, institutions and organizations that contribute to the maintenance of that trajectory on a path to growth, even after growth has ceased and decline set in. The process can be beset by the problem of 'lock-in' where routines and institutions are so inured to signals regarding the superiority of system trajectories elsewhere, or innovation which drastically undermines the validity of the evolved trajectory, that institutions and organizations become paralysed or, worse, panic-stricken. Strong ties between institutions and organizations link industries to each other and reinforce mutual dependence which become obstacles to change. Long-established and implicit consensus regarding the appropriateness of past arrangements for future actions is hard to re-shape. Economies of this kind or in this condition are clearly not learning economies, except insofar as their actors and organizations continue to learn only from the past.

This is part of the explanation of the puzzle Malthus put before Ricardo as to why some economies are rich and others are poor. The normal condition of economies, it was demonstrated in Chapter 2, is that of disequilibrium, in which some nations or regions have grasped the fundamentals of competitive advantage and its undergirding institutional fabric of cooperative routines for the moment, for a mix of industrial activities. The classical task of industry is to maintain competitive, comparative advantage by whatever legal economic means possible. The task of government is to create conditions, including assistance in dealing with market failure, enforcing legal and property codes and monitoring the overall performance of the economy, taking legitimate actions as appropriate, to protect and safeguard the welfare of its citizens.

Central to the latter imperative is seeking to moderate the disparities between rich and poor economies. Nowadays that responsibility is increasingly devolved beneath the level of national government, except in respect of fiscal, budgetary and monetary policies, and in many instances the room for manoeuvre in these spheres is circumscribed or removed to a higher governance level, as in the European Union. In the days when national economies were sovereign responsibilities of governments, protection of citizens against poverty could, in part, be discharged through enforcing or inducing the movement of jobs to people. Little thought was, or even could be, given to the nuances of cultural variation, social capital, interactive learning and innovation in either recipient or donor regions or countries. This was because firms were larger and more hierarchical, thus capable of meeting many of their requirements in a 'make' rather than 'buy' context, and economic governance was far more demand- than supply-led.

It is mainly with the shift of economic co-ordination and competition from stand-alone to supply-chain, network and cluster-like modes on the part of industry and the shift of economic governance from demand to supply-side factors, that new concerns have come to the fore. Among the most pronounced of these are those that fall under the broad rubric of 'learning economy' characteristics. For firms, learning of good or best practice in respect of innovation, management, organization, marketing and leadership, maybe through 'benchmarking' other firms, has become an imperative. For national governments, but particularly for regional governance systems, that hitherto had a relatively passive role in economic policy, similar concerns have moved to the head of the agenda. The focus on supply-side concerns means these governance regimes and those at the lower city or municipal level looking at themselves with a cool, sceptical eye and, if weak, searching around for role models elsewhere, or, if strong, and learning-minded, looking to foresight for new challenges.

The privileged, strong or successful economies of the twentieth-century *fin de siècle* were thought to combine in judicious ways, competitive and cooperative strengths. Hence, weaker economies and, particularly, the 'policy entrepreneurs' who have responsibilities for improving their supply-side offer, whether through 'recruiting' inward investment, retaining existing investment or promoting business entrepreneurship, seek to learn and reflect upon the perceived supply-side assets of the accomplished settings. In this quest, a number of phases have been gone through: sites and services; development grants; recruitment of inward investment; support for endogenous development; and cluster and network promotion. Rather than being wholly dispensed with, each phase has accreted successive phases so that economic development policy now involves a package which, paradigmatically, might involve recruiting a major inward investment by offering land, tax breaks on plant and equipment, workforce training, assistance with supply-chain development to indigenous SMEs, and, if desired, linkage with a university for high-level training and/or research partnership. Each phase lowers the appropriate governance level towards the local.

It is at this point of beginning to manage the implantation of a new kind of externalized, regional or local, *milieu*, of diverse actors and organizations that the issues of trust, social capital and synergies from interactive learning come to the fore. The learning economy demands grassroots leadership: 'Under this model, corporations invest in regions to gain access to specialized workforces, research and commercialization capacity, innovation networks, and unique business infrastructure' (Henton *et al.*, 1997, p. 9). Firms themselves are only capable of generating such *milieux* by themselves on occasion. More often, partnership between firms, associations and governance mechanisms seeks to substitute for the organic evolution of central Italy or Silicon Valley by acts of policy. Among those acts of policy nowadays are efforts to build social capital through initiatives that seek to capitalize on the synergetic possibilities of inter-firm, inter-industry and international relationships.

We shall see in Chapter 5 what kinds of initiatives are being taken to create localized and regionalized learning economies, but it is safe to say that all of them seek to induce or make apparent trustful relationships among economic actors of various kinds. We have seen in this chapter how trust is by no means a monolith. Firms, in particular, are known for under-valuing trust in economic relations, but research regularly shows that firms listen to, and learn most from, other firms (see, for example Cooke *et al.*, 2000). However, this implies that firms are willing to be receptive to degrees of Competence Trust which may be followed up with Contractual Trust and even, in time, Goodwill Trust. Here lie the origins of social capital with an economic face, though as was discussed, whether or not it has direct social, cultural foundations as well is an, as yet, under-researched question. Leadership can involve a degree of risk-taking, as can the willingness to be trustful and follow a trajectory which may or may not lead to synergetic relationship that express activation of social capital.

It was concluded that actors with innovative intent, engaged in the exchange of tacit knowledge, often requiring iterations that benefit from geographical proximity, are most likely to be socio-economically interactive. Innovation is, of course, tightly intertwined with learning for purposes of commercial proto-typing, development and commercialization. It is no surprise, therefore, that innovative action is conceived, intellectually, in systemic terms. The most accomplished economies have in common this systemic information and knowl-edge flow among partners of consequence to the realization of innovation (Lundvall, 1992; Nelson, 1993; Freeman, 1995; Edquist, 1997; Braczyk *et al.*, 1998). The evolution of innovative capacity in a context of legal and property encoding and legitimacy relies fundamentally on the activation of social capital and the propagation of trustful relations among diverse actors. Accordingly, it is an important element in the explanation of economic disequilibrium and the persistence of rich and poor economies. Innovation is not the only driver of economic growth, but, although economists disagree about just how important it is to productivity growth, all are agreed that its contribution, appropriately defined, is likely to be greater than any other factor of competitive advantage.

So, we return to an assertion made at the outset of this chapter, that in a Knowledge Economy the capacity to learn, institutionally and individually, is a key imperative. Learning is profoundly socially interactive. Social interaction is massively facilitated by proximity, though certain kinds of documented or codi-fied knowledge can be successfully learnt at a distance. Trust is a fundamental requirement and, if successful, outcome of learning processes. Trust involves free exchange, confidence and goodwill. As such, it is a key source of social capital, which as we have seen, is a highly prized intangible asset of many accomplished economies. It is a necessary, though not sufficient, answer to Malthus' question, for out of trust and reciprocity come collective action that can overcome the dismal, unreflective conclusion that, in a situation where one has a choice of reliance upon another, one should 'always defect'.

5 Networks and clusters in the learning economy

Introduction

We have seen how trust and learning are vital elements of social capital when activated in the practices of firms, especially those involved in externalized supply-chain relationships. Trust is the reputational 'glue' which holds the 'pieces' together such that suppliers, customers or joint product or service developers perceive each other to be 'as good as their word'. Firms may move from a position of placing 'swift trust' in a partner by making small initial commitments and learning over time whether trustworthiness is sufficient to risk, further, large commitments. From this, through 'slow trust', the kind of goodwill externalities discussed in the previous chapter may arise (Lazaric and Lorenz, 1998; Sako, 1992). Trust and learning go together but they are not interdependent. Learning can very easily be envisaged occurring from a low or no trust context of the 'cry wolf' or 'once bitten, twice shy' variety. Nor need this be only negative learning since it can help second-guessing, a useful piece in the armoury protecting risk-taking.

Thus, learning has a broader compass than being solely trust-dependent, and trust is not necessary for all kinds of learning, but, argue Lazaric and Lorenz (1998), it is a necessary condition for *organizational* learning. Thus while trust promotes the possibility of learning in many contexts, individual learning can occur in its absence. But in the collective learning context, such as that of a firm or an organization, trust is much more important. In modern business terms, this was poorly understood until very recently. Thus the organizational learning literature of the 1970s and 1980s (Argyris and Schon, 1978; de Geus, 1988) made little reference to trust, or the lack of it, as a source of learning difficulties in firms. For de Geus (1988) the task in hand was focused more on how to turn an organization used to learning established rules into one that could rapidly learn new rules, moving from learning by assimilation to learning by accommodation. By the publication of de Geus (1997) *trust*, though still not at the heart of his analysis of the role of learning in firm longevity, was beginning to be projected as an issue for management. This was intertwined with the move, in modern organizations, from a Taylorist, hierarchical managerial mode, influenced by military organization, towards a more decentralized, empowering and devolved style. In a telling illustration of this, revealing a lack of trust or even serious thinking about trust in employees who remain after a downsizing, the case of Exxon is instructive:

> Exxon let 15,000 people go in 1986, in the wake of the oil-price collapse. It concentrated power in a narrow chain of command and took away one side of its organizational matrix structure. In the process it considerably reduced its managerial capacity. A year later [sic], the *Valdez* oil spill took place. It took the company 48 hours to react. That 48 hours has cost it $3 billion so far in cleanup costs, bad publicity and legal fees. And the ticker is still counting.
>
> (de Geus, 1997, p. 155)

Even though Exxon has deep pockets, that is still an expensive way to learn the importance of 'institutional memory' and trust not command as a management imperative both inside and outside a firm.

In this chapter, we will be mainly concerned with examples of, and efforts to engender, social capital between firms, though, where appropriate, reference will also be made to intra-firm practices to enhance interaction. From a policy perspective, a key concern is the extent to which, if social capital is a missing ingredient of economic development, it can be accelerated, induced or forged in unpropitious circumstances. We shall be particularly concerned with two closely related but distinct modes by which firms enter and, in many instances, remain for a considerable time, co-operators as well as competitors through their involvement in business *networks* or *clusters*. The two are not the same and space will be devoted to differentiating them. Importantly, though, they are both economic forms of social capital in which trust and learning are centrally involved. More than that, because of the ways in which economic success has often been assigned to the presence of inter-firm networks or clusters in countries or regions, a premium has been placed on policy-making that seeks to learn from best-practice. Lessons may then influence programmes in support of networking or clustering in less successful economic environments. In what follows we will, first, aim to answer the question as to what these interactive forms are and how they manifest themselves in reality. Then, a review of programmes that seek to support firms by encouraging interaction of one or other kind will be provided. Finally, a summary is given of key points arising from evaluations of such attempts to activate social capital through support programmes, along with a set of concluding comments regarding what has been learned from these exercises.

Networks and clusters: in what sense social capital?

It will be recalled that the previous chapter concluded by saying that social capital is worthy of exploration as an economic instrument even though its conceptual origin lies in the sociological analysis of civil society. This is notwithstanding its early usage by Jacobs (1961) in her study of the social and economic life of cities, and subsequent analyses of modern forms of economic organization (Jacobs, 2000). This is reinforced by Putnam's (1993) observation that 'social capital is coming to be seen as vital in economic development around the world', a theme taken up by authors associated with advising on economic development policy (Henton *et al.*, 1997). The work in question, evolved from advisory work by SRI, the Stanford research consultancy, places emphasis upon social capital building on revitalizing distressed economic areas and maintaining the competitiveness of areas vulnerable to cyclical shifts in markets. The particular form taken by social capital is that of 'economic communities' based on associative governance practices by private actors exercising leadership while also gaining support from public stakeholders for an economic development 'vision' (on this, see also Cooke and Morgan, 1998). It is probably gratuitous to add that the world of networks and clusters is remarkably popular with both consultants and policy-makers not least because of their promised advantages for firms and policy-delivery agencies based on the simple message that cooperation pays dividends at relatively modest cost.

The 'economic communities' analysed by Henton and colleagues are typified by the ways they use social cohesion mechanisms for economic ends. They pursue regional economic development through practising collaborative strategies articulated by a new kind of leadership called the 'civic entrepreneur'. Strategies are founded upon linkage and interaction among business, government and community leadership. Civic entrepreneurs are the catalysts for creating regional economy and community relationships, operating across boundaries. Such leaders are most likely not to be traditional politicians but rather business or community actors with strong and widespread personal networks among diverse constituencies. The reason advanced by the authors for why they have risen to the fore in the USA is globalization, which closes off older economic development avenues as it opens up others. Information technology offers communities that were hitherto by-passed or deprived of their living a chance to compete by linking themselves to global firms and marketplaces: 'The shift will be fundamental – from the centralized, vertically integrated, model of business and government ... toward a more decentralized, horizontal, and networked regional model' (Henton *et al.* 1997, p. 9). This echoes almost perfectly the observations on intra-corporate restructuring of de Geus (1997), except it refers to the extra-corporate world of multifarious, variably sized economic units and organizations. While some examples of such economies are, like Silicon Valley and Austin, Texas, paradigmatic 'new economy' locations, others such as Cleveland, Ohio are classic 'rustbelt' cities that have been revitalized by associative activity.

Networked regions are seen to be the fashionable ones for footloose corporations to seek out where they integrate high class design, manufacturing research and marketing rather than offering low cost land and labour. Large corporations seek to integrate with such regional milieux, develop roots and operate globally from a trusted and secure base with specific, scarce assets. So a key policy task is to understand the regional economy, its strengths and weaknesses, and stimulate an ecology favourable to high skills, high value-added and high-income jobs. These settings are likely to be far more specialized than those to which the old division of labour based on the pursuit of low-cost locations for routine assembly work gave rise. Whereas the latter became specialized mainly in the modest skills of their 'peripheral Fordist' workforces, the former may specialize as innovation regions, financial services specialists, contract manufacturing areas, or specialist industry clusters with upstream and downstream competences.

Why the emphasis on local and regional interconnectiveness in an economy where global forces seem to have triumphed? Because the network model of regional economic development demands that the 'exploded' production system, now externalized to the region rather than internalized to the corporation, has the ability to reintegrate itself, especially for crucial services or components that are the core competences of complementary firms. It is through the network that leadtimes are reduced and time-to-market of novel products or services is enhanced. In a recent book, Porter (1998) demonstrated how, on three key indicators of competitiveness – productivity, innovation and new firm formation – clustered production systems are superior to hierarchical corporations. The explanation is that knowledge, human capital and technological applications flow more swiftly because of the 'open systems architecture' of the cluster compared to the silo-like vertical channels of the hierarchical corporation. This is even true of the learning organizations that de Geus (1997) writes about for whom scale, technological trajectories, and corporate culture, among other factors, can remain barriers to agile identification and exploitation of opportunities for innovation and productivity gains. In electronics, for instance, industry leader-firms often recognize this and simply acquire innovative start-up firms, the strategy of Cisco Systems, or develop *keiretsus* of favoured start-ups funded through corporate venturing as with Intel.

As speed rests on communication of vital information concerning, say, customization or late configuration of products or services, knowledge that may demand several iterations to realize, proximity offers an added advantage over distance. This is why firms that themselves grew quickly from the start-up phase choose to remain in the technological milieu from which they sprang. Hence, networked and clustered models of regional economies offer irresistible attractions both to firms because of the 'spillovers' (Audretsch and Feldman, 1996) that provide opportunities for expansion, and to policy authorities because of the success that attends firms establishing in such settings. But because they are judged by different performance indicators and thus will pursue different policy agendas left to their own devices, the network model offers a compromise or third way between market and hierarchy criteria conforming to a non-market, non-hierarchical, mode of policy formulation, the catalyst for which is the 'association' led by a civic entrepreneur.

Thus, conceptually at least, social capital for economic purposes is seen to be more comfortably realized in contexts where the antagonisms of 'normal politics' and the special interests and special pleading of business are removed from the role of civic entrepreneurship. It is a question of trust. Trust found in networks and clusters is volunteered by members who understand the potential for mutual advantage to be obtained from it. Similarly, work for the economic community is best when volunteered or at least not-for-profit. 'Community' is the 'sacred institution' that demands the loyalty of associational actors, not the partisanship of special interest groupings. To the extent such associations gain legitimacy, they may be rewarded with modestly devolved budgets and powers and because they are composed of networks of contacts with resources of their own to allocate, perhaps on the basis of a certain amount of 'guilt-tripping', such organizations may be able to act, solve problems or see opportunities quicker and clearer than the humble committee-laden bureaucracies or management structures of individual public administrations or firms.

We should be clear that, Osborne and Gaebler (1992) notwithstanding, this is not a description of government re-invented, although there are some features in common with their more wide-ranging prescriptions. Their aim has been to encourage learning by government of some of the management expertise of good companies across the board of government obligations. Thus in the name of better value for taxpayers' money, the option was in some cases exercised of privatization of all but the town council's members who, to preserve the fundamentals of democracy, are voted in but may meet only occasionally to decide on the dispensation of contracts. Whether the risk of private firms failing to adhere to legal health and safety or pollution regulations is safeguarded in this kind of approach is an open question, and one which some would say is becoming more clearly answered in the negative by the experiences of some radically liberalized Californian governments. But in discussing economic community building, Henton *et al.* (1997) are painting on a far narrower canvas concerning that area of community life over which local and regional governances have least control and influence, namely, the economy. The networked economy, in particular, creates its own governance demands, which are difficult for traditional government (i.e. the public dimension of governance) to understand, let alone fulfil. Hence, this is a response to uncertainty from actors of consequence to the local or regional economic community to improve the collective competitiveness of firms within it by improving openness and communication, raising learning capability and, through promoting collaborative practices, building up trust and social capital.

This is the theory, which Henton's team are among the first to have articulated as a role for civil society which does not usurp that of democratic politics. It is more than traditional civic boosterism of the kind well-tried and tested in American cities and beyond, not least because it is not simply an agreement by government to give money but otherwise step out of the way of big business' efforts to sell and develop land for speculative office and leisure developments. That only slightly overdrawn caricature is rightly criticized for its disposition in most cases to offer a public subsidy for private profiteering. The civic entrepreneurship approach is more firmly rooted in networking philosophy, which is inclusive, diverse in its membership and functionally directed towards skills development and educational investment as well as business enhancement to improve the ultimate competitiveness of the regional or local economy in tune with an 'all boats are lifted by the rising tide' mentality. It is unclear whether, in practice, this is how the approach works out or whether life is more egalitarian in places that provide the exemplars. The consultants and policy-transfer agents, whose task has been to persuade others that social capital can be activated as it has been in the originating countries and regions, are persuaded that it works. The best work on this from Italy (Trigilia 1992) suggests they may be right, but it is too soon to say in the USA.

Networks and their discontents: passive to active social capital

We have traced the development of conceptions of trust and learning, conceived as key elements of social capital and seen as the missing ingredient in explaining economic backwardness, from recognition as an ingredient in the success of networked economies, to being advocated as aspects of civic entrepreneurship for boosting economic communities. But we need to step back somewhat from this transition because it runs the risk of introducing a major paradox. The paradox is this; networks and clusters are rather like the Internet in that nobody controls them but their users. But when their perceived success factors are distilled into their essentials and transformed into a policy instrument, they come under the control of civic entrepreneurs and Henton's 'grassroots leaders'. We shall see, towards the end of this chapter, how other policy models evolved from the same template also gave rise to disappointment on the part of their initiators, particularly because control of the processes stimulated into existence could not be assured. This is far from an argument against adaptation but it is an observation that the kind of economy we are now witnessing cannot be induced or nurtured satisfactorily by a policy-making and policy-evaluating machinery which is not itself more interactive, agile and flexible in form.

Let us begin by exploring reasons why this should be the case and use that discussion to lead into an appreciation as to why a specific phenomenon, the *industrial district*, widely understood to be the progenitor of both networking and clustering as conceptual and industrial policy models, should have become so influential as a policy concept throughout the world. Unlike financial capital and human capital, social capital is not appropriable and legally capable of protection. If it is 'owned' at all, it is the property of those who are party to its expression and activation. It degrades as relationships between those in the partnership or network degrade, although it can be revived if there has been no breach of trust, since it is understood that certain key social capital assets may be more valuable at some times than others. Authors such as Burt (1992) suggest, perhaps exaggeratedly, that social capital determines competitive success because it is the knowledge key employees in a firm have of customers and the market more generally that determines the profitability of a product.

However, it is worth remembering that if the quality or cost of a product is inadequate, social capital is likely to degrade rapidly. Nevertheless, the notion that social capital is valuable and non-appropriable but nevertheless constrained by trust helps understanding of its attractiveness as a potential economic development instrument, to the extent it can be, as it were, bottled.

The most important elements for economic purposes are twofold: sources of useful information capable of adding value, or what we may call 'knowledge capital' to differentiate it from information of a general nature that has not been screened into more and less useful categories, and reliable means by which such knowledge flows and enables the recipient to learn things that are of consequence to adding value. We may call this second asset 'learning capital'. Knowledge and learning capital differ in that while knowledge capital is a given, albeit with value-adding attributes, learning capital is a cumulative process that results in a narrative or theory being capable of composition that involves creation of new knowledge. This is consistent with two points made by Schuller and Field (1998) about the nature of learning as a collective process in relation to social capital. First, they point to the implicit or tacit nature of knowledge associated with social capital and contrast it with the codified, or transferable skills associated with human capital. Social capital, they suggest:

> refers rather to the ways in which diverse areas of knowledge, or skills, are pieced together by more than one person, not necessarily operating at the same level but complementing each other at least to the extent which makes forms of learning possible which would not otherwise have been so.
>
> (1998, p. 231)

Second, they examine the conditions under which social capital can both help and hinder learning. The key to this is trust. Thus, referring to social capital and learning in Northern Ireland, they find high social capital (associationism, etc.) and high educational attainment. But participation in post-school, continuing education is low because employees draw on personal networks (presumably sectarian, at least in part) for career enhancement. Nevertheless, they conclude that the apparently strong presence of social capital enables both formal and informal learning to be better than it would otherwise be and that while:

> formal learning is associated with the development of routinized systems for transmitting and recognising knowledge and skills, less formal types of learning are promoted by the pooling of information and sharing of capacities which arise when levels of trust, reciprocity and common norms are high.
>
> (ibid., p. 233)

Even though they point to the non-existence of even 'weak ties' as discussed by Granovetter (1973) between some of these personal networks and also comment on the popular view that interests may make proximity redundant, their conclusion is that *proximity* is important to the successful, practical operation of learning processes where trustful interactions give rise to both the normative institutions and asssociational organizations which comprise social capital.

Ideas such as these have been expressed in different but cognate terms about spatially concentrated industrial districts in northern and central Italy. Their characteristic features, strengths and weaknesses have been analysed by experts such as Brusco *et al.* (1996), Belussi (1996) and Varaldo and Ferrucci (1996). The study of the district phenomenon is well established, and books such as those edited by Goodman *et al.* (1989), Pyke (Pyke *et al.*, 1990; Pyke and Sengenberger, 1992) and Cossentino *et al.* (1996) give good general and specific accounts. But few of these explore very profoundly the nature of social capital, trust and learning in the networks that underpin the districts (though Dei Ottati, 1994 and 1996 are an exception). What is now most needed is to view these phenomena in evolutionary terms and this the first three sets of authors cited above, begin to do. Brusco *et al.* (1996) show that, on a number of key variables, firms in districts, characterized by specialized, geographically concentrated sub-contracting relationships within a specific industry, outperform the industry in general. Examples include food processing at Parma, motor-cycle production at Bologna, ceramic tiles at Sassuolo, shoe manufacture at San Mauro Pascoli and garment manufacture at Carpi – all in the Italian region of Emilia-Romagna. Here earnings, employment and exports grew between 1984 and 1993 more than the average for these industries at the national level. This was not true of other districts such as Reggio Emilia, specializing in farm machinery, and Piacenza in machine tools, but a majority of firms in districts performed better than the overall industrial sector of which they formed a part.

The evolutionary questions are why, and where, in future, developments may go. Brusco and colleagues conclude that the answer to the first question is the predominance of small firms in mainly informal network relationships with others and with enterprise support organizations in the absence of strong governance mechanisms. Self-governance through learning of, for example, the local implications of globalization or the acquisition intentions of large firms and responding through internally organized concentration processes, such as the formation of groups of firms is considered to have produced an appropriate response to market changes without the need to introduce a 'central control unit' to manage district policy. Regarding the future of such districts, conclusions are positive but cautious. Because the network form of the industries is so efficient at transmitting knowledge capital around the system, more open than markets – which encourage 'back-to-back' as well as face-to-face interactions – and, more flexible than a central control unit, the demand to adapt to global integration and competitiveness will 'select' those forms of inter-firm arrangement that have best learned the evolving conventions and rules of the game. Thus, decentralization to Asia and Eastern Europe, extending the network for lower value-adding activities, is thought unlikely to weaken the 'district' form of local production systems if strategic, high value-adding functions like management, innovation and marketing remain. The evolutionary trend is thus towards districts characterized by knowledge-intensive services within a specific industry. But innovation may also mean advanced production activity may remain in or even return to home base.

The perspectives of Varaldo and Ferrucci (1996) and Belussi (1996) are less sanguine as to whether the evolution of the districts to higher value-added and qualification levels is as likely as their compatriots think. The first of these contributions explains the 'why?' of apparent industrial success in terms of Marshall's (1919) insight regarding the importance of economies or, in modern parlance, 'spillovers' external to the firm but internal to the district or sector in which it exists. These include concepts such those discussed earlier, notably collective, complementary learning opportunities and processes. Such learning is facilitated where size, sector and space of location inter-connect, because when there are many small firms, variety and variability are present, enabling a wide range of responses to 'ecological' challenges to the system. These might include short-term increases in demand for manufacturing capacity, differentiation of orders, customization and even late configuration requirements, and so on. Further, transactions costs are reduced by word of mouth agreements ('untraded interdependencies' as Dosi (1988) terms them and Storper (1995) adapts them to network contexts), innovation of an incremental nature is facilitated by rapid circulation of information and learning, one consequence of which is that productivity levels are higher than average, noted also more generally by Porter (1998). Investments tend to converge towards common technical standards, minimizing the difficulties caused by incompatibilities between technical systems, and organizational innovations associated with common equipment can be rapidly diffused. On-the-job learning means that skills are abundant and relatively cost-free to provide to workers, and understanding of industry needs means skills are adaptable not only to production in or servicing of the core industry but also to production of equipment and specialist services needed in the industry. There is a widespread understanding of entrepreneurship, social distance between employer and employee is low and positions are quite capable of undergoing reversal over time. This describes a thoroughly 'embedded' economic setting with high trust, collective learning and social capital strongly evident.

However, as Varaldo and Ferrucci (1996) argue, it is vulnerable. Its strengths can also become its weaknesses as both model and reality confront external stresses and change. Shared understandings and deep knowledge of local business processes and practices learned 'organically' on the job rather than 'mechanically' in the classroom mean entrepreneurs are averse to introducing new skills and conducting organizational innovation. New competences are not welcome where they differ significantly from the owner's practical knowledge. Local sub-contracting reduces the pool of possible suppliers and may, thereby reduce efficiency. Diversification, except in relation to the capability of district firms to evolve their specialization in a specific part of the division of labour, is not welcomed nor seriously considered. Localized social networks, as in

Northern Ireland, may entail certain externalization diseconomies which make firms reluctant to consider changing location, labour or linkage patterns to more profitable ones. In other words 'lock-in' to a specific evolutionary path has been both the advantage of the industrial district, especially lock-in of the first-comer variety enabling the appropriation of super-profits for a period but, when the environment changes, lock-in becomes disadvantageous unless change can be responded to collectively rather than individualistically.

The analysis here is that globalization and innovation pose the greatest challenges where economic communities display lock-in and path-dependence, and that firms as members of collectivities must be flexible to the nuances of these forces as they evolve in future. Collective learning, whereby the districts as collectivities can exploit the advantages of both forces while seeking to protect themselves from threats, must also become embedded. Closer linkages with distributors active in global markets and innovators absorbing global best practice are called for. Evidence from Veneto region by Cappellin (1998) shows one novel response to this.

> An interesting phenomenon is the increased activity of many small entrepreneurs originating from the same region in the same foreign countries, such as textile entrepreneurs of Veneto region in Romania ... [who] have undertaken tightly joint investments aiming to create artificially a sort of 'industrial district' ... it is similar to a gradual 'learning process' ... to the different environment.
>
> (1998, p. 9)

This is clearly comprehensible in terms of an evolutionary process of reproduction through learning and adaptation to a new ecological situation.

If we step outside the 'old economy' sectors that are characteristic of Italian industrial districts and apply this problem to 'new economy' networks as found in California, we find significantly less problems with lock-in. The obvious question is why? There are three features of such places as Silicon Valley which immediately suggest key differences. First, social capital is more extensive and intensive. It is more extensive in reaching across a wider span of stakeholders, including basic research science as the knowledge capital which is the source of raw value at one pole and venture capital at the other. It is more intensive because many participants are, on the one hand, rootless recent arrivals, who seek out economic communities within which to swiftly build trustful economic and social relationships. Second, there is more absorbed learning capital expressed in new economy clusters in the sense that, according to Bronson (1999), engineers nowadays habitually understand and exploit (not least through day-trading) knowledge they have learned about venture capital, initial public offerings and stock markets from investors with whom they regularly interact. By the same token, venture capitalists understand technologies far more than their forebears and also gain significant advantage from so doing. In other words, the *complementarities* that characterize the division of labour in a traditional territorial production system have been transformed into knowledge *coalescences* in a knowledge-driven one. Finally, knowledge is constantly being renewed rather than becoming settled and conventional. Tacit knowledge, as yet uncodified, or an idea which can swiftly be appropriated, protected and provide 'first-mover advantage' before being imitated, is the key source of value in contexts which bear all the hallmarks of Schumpeterian 'swarming' around first-mover innovation.

Belussi's (1996) analysis of industrial districts bears comparison only in that she envisages the evolution of industrial districts involving the differentiation of types according to relative sophistication and knowledge-intensity of the industry in question. Thus, some local production systems have global competitive advantage and are not yet visibly likely to be challenged, and may retain significant global competitiveness by virtue of their control and influence over the global networks implicated in that industry. Other sectors may have strength in global and lesser markets but be more in thrall to more powerful distribution companies, retail chains and the like. Yet others, characterized by traditional mentalities may lose ground internationally to competitors but retain a worthwhile presence in regional or national markets, especially where, as with clothing or furniture there is a strong loyalty-effect shown towards indigenous design expertise. In each case, however, 'territorial embeddedness' of the kind discussed, built on established social capital, means that districts of this kind can survive and develop provided that they are not over-governed and that expressed needs are met by whichever provider is the most appropriate with 'new policy' only being a last resort, designed in collaboration with its recipients. Then the collective and interactive learning proficiency may still be retained even if, as all these authors predict, 'leadership' of the 'economic community' may have already become a key feature as firms co-ordinate their activities in groups to deal with the pressures of globalization and innovation-based competition.

Returning briefly to Californian 'industrial districts', it is noteworthy that a version of group formation has occurred there also, later than in Italy and influenced more by Japanese business practice. We noted how firms like Cisco Systems and Intel have developed partly internal, partly external families or *keiretsus* of suppliers whom they have acquired or in whom they have sizeable equity shares. However, this practice is equally, if not more pronounced among venture capitalists. One of the longest established of these, Kleiner Perkins has its own *keiretsu* of firms in which it has invested numbering some two hundred firms. It is company policy to encourage these firms to engage in inter-trading which is of advantage to the firms themselves since they have privileged access to the innovations of the leading firms, and of advantage to venture capital in that it makes their investments more productive as a consequence.

Thus, within the major cluster of new economy businesses that constitutes Silicon Valley are networks of clan-like communities of firms gaining advantage from high-trust, reputational social capital transformed into a powerful instrument of economic competition.

Network forms of business organization in Italian industrial districts are shown to be flexible, sensitive to cultural nuances, incrementally innovative, responsive to learning opportunities, communicative, efficient, trustful and expert. But if not guarded against they can become introspective, unresponsive, complacent and with emerging uncompetitiveness, distrustful, inefficient and backward. Networks need constant replenishment if they are not to degrade. Their social capital has constantly to be activated in some form, not necessarily or at all times economically, but socially, culturally and politically. This hints at how difficult it is to recreate such phenomena in new settings, even though we hear of many thousands of Italian entrepreneurs doing so in the Latin culture, or despite being overlain with Soviet-style governance for decades, in Romania. But that they exist in industrial district form in many different national and regional settings at a variety of levels of economic development is testified to by the work of Pyke (1998) who reviewed literature showing their existence in countries as diverse as the USA, Pakistan, Indonesia, Ghana, Zimbabwe, Brazil, Peru, Mexico, Ecuador and India. Interestingly though, they are not found evenly distributed in such countries, but localized in particular geographical settings where, presumably, social capital is more pronounced in its economic dimension than elsewhere.

Findings such as these, emphasizing social capital and its economic expression in specific localities or regions, echo many of those discovered by Scandinavian researchers approaching the question of networks from a business studies and marketing background. These were neither exploring social capital nor spatial location as economic development factors. This literature, represented by writers such as Johannison (1987), Hakansson (1987) and Johanson and Mattson (1985) is less far-reaching than the networks literature discussed thus far, yet these authors found that marketing by firms of their products and services did not simply occur through arm's-length exchange but, rather as Burt (1992) has subsequently argued, by the exercise of social capital through networks in which trust, reputation and reciprocity are key. Markets are thus, themselves, far from the competitive arenas of individualistic utility-maximization presented in neoclassical economics and much more influenced by social exchange which evolves from small-step, incremental trust-building which is low risk in leading to more committed agreements. In the process, partners adjust their expectations and understandings in ways that minimize disturbances to the risk–trust equilibrium.

It could be said that the Scandinavian approach ultimately rests on a version of neoclassical equilibrium theory even though it rejects the economism normally associated with that perspective. The question arises as to what would happen to the cosy arrangements described in the event of a shock to the system such as recession or a more exacting competitiveness environment brought on by globalization tendencies in markets. Are customer–supplier relationships reflexively nurtured to adapt and accommodate to the new conditions and what if traditionally supplied products or services are no longer up to standard? Do trustful relationships triumph over narrow bottom-line criteria? If so, then this literature is strengthening considerably the finding by Brusco *et al.* (1996) that firms perform better in networks than in markets. The argument put, which suggests this is the case, is that finding another supplier, given the social capital already invested in the existing supplier (or customer), which would also have to be reinvested in a replacement, makes defection expensive in time, energy and money terms (Johanson, 1991). Defection only works on the 'spot-market' for suppliers or customers, or at the stage when partners are feeling their way towards an implicit or explicit contract. That is, before embedded social capital has been built up.

Perhaps the most interesting kinds of networks are those formed for the purposes of advancing innovation within partner-firms. Hakansson (1989) has written extensively on this, showing that user–producer linkages focused on innovation are the most durable kinds of inter-firm cooperation, averaging ten years. The key reasons why this should be so are; continuous learning opportunities from the implementation by the user of the producer's product or process upgrades, and 'lock-in' effects arising from customization. Similar conclusions are drawn by Johanson (1991) who further saw the centrality of networks to producers of innovations since new users are a source of 'link-profits' based on customization of a core technology or application with relatively modest adaptation. From these learning interactions, built-up over time, comes the knowledge-capital to develop the next wave of generic core technologies or applications. Part of this process also involves a global know-how scanning function, which helps overcome the constraints of localized learning. This, as we have seen, now exercises commentators on and practitioners within industrial district settings. The two bodies of literature thus come together in agreement on the importance of what Grabher (1993) refers to as 'systemic loose-coupling' whereby the weaknesses also implicit in the 'strong ties' centred upon localized learning are offset by the strength of weaker ties to a population of other networks operating in regional, national or global space.

Networks, learning and policy transfer

Thus, we arrive at a portrayal of networks and their characteristic strengths and weaknesses in the field of inter-firm interaction, innovation and economic development that is as follows. First, they are relatively small scale in terms of numbers of firms involved. Thus, even though in some industrial districts there may be thousands of small businesses, they do not all constantly interact with each other. Rather, as we have seen, preferred cooperators of supplier firms are the most formally linked, evolving lately into groups though, informally, information may flow between larger numbers of inter-firm relationships. We can thus say, second, that networks have restricted membership and, further, that this often takes the form of horizontal relationships between firms of comparable size rather than those more typical of a supply-chain hierarchy to a large firm. Third, we can say that networks rely on the evolution of strong links derived from trust, reputation or reciprocity. The social capital involved means that although firms engage in networking to become stronger, more competitive actors, they do so by using cooperation as a competitive weapon towards other firms outside the formal network arrangement. Fourth, networks in the proper sense are formalized implicitly or explicitly. There may not be a legal contract underpinning the network but there is likely to be some form of implicit contract even if verbal or handshake-based only, and agreement to take promised actions, commit a given amount of resources, or otherwise be reliable. Finally, networks imply a high degree of agreement about the pursuit of common objectives if they are to be successful, and a vulnerability to low commitment if objectives and designed actions are not clear and acted upon, as well as to dissensus occurring among network members.

An outstanding example of networking of the kind conceived of above, that is, learnt, analysed, transferred and systematized as policy models, then implemented, assessed and revised via an internalized monitoring and learning process would have to be that of the Danish Network Programme. This began when American consultants noted how, in the late 1970s, Italian industrial districts were outperforming large firms in quality markets for traditional goods. The fact that these comprised economic communities of smaller firms cooperating through vertical and horizontal networks, linked by marketing specialists to global markets through sales to large European and American chain and department stores was interesting to economic development agencies in states where jobs were beginning to be eroded in large firms and small firms were seen as a possible salvation. Hence, Hatch (1988), one of the consultants in question, hawked the network model around numerous states, some of which, as in the Michigan Manufacturing Initiative, and later the Pennsylvania Ben Franklin Partnership experimented with aspects of the networking model. An American-trained executive of the Danish Technological Institute (DTI) heard a presentation on this and invited Hatch to come to head office near Copenhagen to meet the DTI Board.

At the time DTI was part of the Danish Ministry of Industry but it was about to be privatized. Marketing a new model for supporting small business growth was seen as a good business opportunity and one that could help sustain the fledgling DTI company in a new business service area. DTI worked on the networking concept, analysed the key elements of the success of the organically-grown Italian system, systematized these into an action programme and sought the support and financing of the Ministry of Industry to implement a Business Networking Programme for Denmark. In 1989 the ministry agreed to set up such a programme, recognizing it as a development of traditional DTI business support in assisting cooperatives and trade associations to form as well as selling technology assistance to firms. The programme was set up in 1989 with a three-year budget of $25 million. On the input side the programme had three key elements. The first was a well-funded and executed marketing and promotion campaign, advising firms of the advantages of establishing networks. The importance of this lay in the fact that, despite Denmark's cooperative and associative traditions, entrepreneurs were individualistic, even hostile to a notion of collaboration, unlike their Italian counterparts. So winning support from a sceptical small firm business leadership had to be a first step. This was aided somewhat by analyses of the probable difficulties for small firms in Denmark in facing competition within Europe once the Single Market legislation came into effect in 1993. These showed that Danish small firms were losing market share at home and not performing well in general in export markets because of less than adequate innovativeness, technological investment, marketing and organizational learning (Lundvall and Johnson, 1994). Given the dearth of large firms in Denmark, this had serious potential repercussions for the economy as a whole, hence the interest and generous investment of resources by the government.

Second, a major portion of the budget was to be devoted to a programme of training the key agents who were to be responsible for implementing the programme. These were called network brokers and their task was to create networks at local level by using their contacts and intelligence to persuade individual entrepreneurs to join with others in a network of five to ten other firms. However, it was recognized that this task was likely to be fraught with difficulties that would only be partly moderated by the provision of incentives. Thus, a methodology was formulated for inculcating understanding that brokers were not meant to appear as technical fixers or problem solvers with a set recipe of solutions. Rather, they were to learn to be people who were good at listening and being receptive to the needs, concerns and anxieties of candidate network members while developing the diplomacy skills to help them reach consensus on why networking was relevant and what it would enable them to achieve if they joined. This was not easily achieved, not least because the pool from which brokers were selected was predominantly composed of professional consultants, engineers and lawyers who, according to a programme evaluation done in 1991 (Gelsing and Knop, 1991), found it hard fully to override their professional training with the brokerage principles.

The final key element in the programme was the provision of subsidies in stages to support network building. The programme was marketed actively to a target number of 7,000 small and medium-sized enterprises of which 2,500 had joined networks by 1991. The number of networks initially achieved was 300, of which 175 survived until the end of the first stage of the programme in 1992. Hence the average network consisted of some eight firms plus various non-firm organizations such as university research centres. The subsidy structure can be illustrated by reference to a second stage Tourism Network Programme. This involved training twenty-four brokers. The budget was $10 million and grants were paid to networks in three phases. At the formation of a network, $15,000 or 75 per cent of costs were available to support feasibility studies into the agreed action lines developed by the network. If accepted by the programme assessors, a second stage planning grant of $10,000 or 50 per cent was available for production of a business plan. If that was formulated and approved, a grant of 50 per cent of costs of implementation was available in the first year, and 30 per cent for the second. Hence, these networks were of the formal or 'hard' variety with contractually binding legal agreements, time scales, and agreed action lines to achieve clear business plan objectives.

When the first phase of the Network Programme was evaluated, the results were modestly positive. One of the results receiving a high level of attention was that by 1993 60 per cent of Danish exports to the main market of Germany came from companies in the Network Programme. More prosaically, some 42 per cent of firms in networks ascribed to the programme increased turnover of at least 4 per cent per year and 20 per cent ascribed it a 10 per cent turnover growth between joining in 1989 and being evaluated in 1991. So a clear majority had worthwhile gains to show from involvement, though more than a third did not benefit substantially in bottom-line terms. However, a further and tougher assessment of the programme conducted in 1996 deemed the programme a failure in policy terms, a judgement which caused DTI to change its future networking support towards assisting the process of spinning-out businesses from university research laboratories, with early customers for advice by 1998 being the University of California at Los Angeles and Berkeley. Why the devastating critique? The independent evaluation was that the programme did not achieve its core objective of setting up and maintaining a large number of networks, suggested to be at least one hundred over the lifetime of the phases of the whole project. However, independent research conducted in 1998 discovered that over one hundred networks continued to operate in Denmark, by then subsidy-free, but that the network members, mainly beneficiaries of the Network Programme were now in networks of their own choosing. The new lesson was thus that learning networking had been a sufficiently valuable experience for eight or nine hundred firms that they were continuing to do it voluntarily. The secondary lesson was that strict objective-oriented project evaluation is capable of producing misleading results (Hughes, 1998).

Among the many examples of reasonably successful networking projects that enabled firms to achieve collectively what they would have been unable to achieve individually are the following three. In Fredrikshavn, Jutland, a network of eight firms formed to produce fishing equipment for export to international markets of a standard beyond the capability of any single firm. Members included a producer of electronic bridge equipment, a metals firm, a producer of fishing gear, a food processor and a shipyard. The network also included a bank, research centre and a holding company to manage the affairs of the network. Rosenfeld (1990) one of the two consultants who 'discovered' the Danish transposition of networking from Italy, notes a different network in nearby Aalborg among fisheries to produce and market health products from the sea, such as fish oil. This group encompasses fishermen, processing companies and a hospital. The third example is a seven-firm network in Salling, Jutland, involved in manufacturing furniture. Each was losing market share, they were encouraged to join the programme and immediately formed a trading company. Key tasks were divided so instead of all doing design, two alone were assigned that responsibility. The remainder then specialized along functional rather than complete product lines. One consequence of re-engineering their business is that market share was recaptured and the trading company now exports high quality furniture to the EU and beyond.

The use of networks by firms has undoubtedly increased, although for firms interviewed on the subject it is often the case that from generic and specific populations of SMEs only around 20 per cent are habitual networkers (Cooke and Hughes, 1999; Cooke and Wills, 1999; Huggins, 2000). Huggins cites some forty academic studies of business networking in the 1990s alone, from a wide variety of developed and less developed countries. Thomas (2000) found in a study of three UK regions that firms co-operating in vertical supply-chain networks had a 10 per cent greater chance of being innovative, and that growth rates are higher for all firms in horizontal networks involving agencies as well as firms. Broadly speaking, European studies find networking stimulates improved performance by SMEs whereas North American studies find less performance effect from economic networks than from those centred upon family and friends. However, in an evaluation of the USNet programme, administered by the National Institute of Standards and Technology (NIST; see also, Storper, 1995), Shapira (1998) concluded that participant firms reported positive net benefits but that, as in the Danish case, promotion goals by the managing body were over-ambitious. However, when firms could engage in special network projects where their own network goals were clear and action lines could be pursued straightforwardly, networks were indispensable for achieving success in projects that demanded a growth-orientation from firms. Asian studies are fairly consistent in stressing the success of networks with a strong local base but global reach, while those from Latin America often show that a network approach is centrally important to the practices of SMEs whatever the market scale at which they seek to operate (see, for example, Schmitz, 1998).

A wide-ranging review of networking among US business firms by Malecki and Tootle (1996) confirmed the general conclusion that whereas European countries and firms in them are well attuned to networking and programmes to support it, those in America are less so, except for innovation, where colleges may be partners, and 'flexible manufacturing networks' which were being urged upon US small business as a response to the perceived superiority of Japanese *keiretsus*. Interestingly, the US case is presented by these authors as one in which informal interactions facilitated by geographic rather than functional proximity are most valued and used by SMEs. Geographic proximity would imply business clusters, to be analysed in the next section, rather than the kind of close integration that might, in principle, be feasible at a distance through the use of e-commerce, e-production, and the like. That the former is more important for regular networking activity in 'old economy' as much as 'new economy' industries is also testified to by Porter (1998) and Saxenian (1994). It may be concluded that in the USA, markets function well and networks rest upon such relations to a considerable extent where business clusters are found. In Europe, by contrast, markets function less freely and public bodies intervene to tackle situations of perceived market failure. Firms are comfortable with this whether proximity is geographical or functional. But tellingly, US business innovation and new firm formation rates are higher, suggesting again one of the barriers to European competitiveness resides in too much dependence on public initiative in creating conditions taken for granted due to superior market functioning in the USA. This is notwithstanding the cited success of networking as promoted by NIST which, it will be recalled, worked best when firms came together for the pursuit of projects in which each had a specific interest which could be fulfilled through co-operation.

Hence, to conclude this section we can say that networks are an established but not necessarily mainstream part of modern business practice. Most multinationals have 'strategic alliances' but they already have a slightly dated air to them and merger and acquisition practices have become pronounced once more. Leading edge research is now much more the province of specialist research consultancies, or in 'new economy' sectors like so-called TMT or technology (biotechnology included), media and telecoms, small start-ups are simply acquired by the likes of Microsoft, Intel and Cisco Systems or supported with milestone payments from big pharmaceuticals firms. But networks exist in many countries and many fields. In numerous countries such as Norway, Denmark, Finland, New Zealand, Australia and the USA, national network programmes have been implemented with some success as a means of assisting SMEs to become globally competitive. Accordingly, it is possible to speak of varieties of networking character and structure:

- Informal networks, rather like those typically found in the USNet programme based on family, friends or business partners.
- Formal networks which are among firms but include financiers, accountants, lawyers and other professional advisers.
- Soft networks are on-going and give actors with a broad common interest occasion to meet and share experiences. This might include innovation, management upgrading or marketing, for example.
- Hard networks are contractual, legally binding and action-oriented with a business plan and set time horizons. Danish networks are often of this kind
- Vertical networks are focused upon supply-chain or supplier club and supplier development group activities. They may be formal and associative and will result in some hard network contracts.
- Lateral networks link firms of a similar size with complementary assets, though some between competitors are known (e.g. Mercedes suppliers in Germany), who generally work in formal or hard networks to achieve some business objective they could not consider individually.

Langlois and Robertson (1995) give a large number of cases of, on the one hand, core networks which are quite similar to vertical networks as discussed above, notably Japanese *keiretsus* or tiers of hierarchically structured supplier relations to a core customer. On the other hand, they identify decentralized networks more common in the West where suppliers, in their examples, are likely to be systems or modular suppliers with a need for compatibility with their own, their customers' and their suppliers' parts and components in a multi-client system. This is more like the Danish networks, producing integrated products from a discrete set of horizontal linkages brought together on a project basis. This form of production of goods and services became prominent in the 1990s, as work on contract project engineering in Germany has shown (Schamp, 2000). If one were to map the evolution in production (of goods and services) over the twentieth century, particularly in terms of its leading paradigms it would resemble the matrix in Figure 5.1. The key features are two evolutionary shifts. In the transverse dimension there was a shift towards network forms of interaction among forms as externalization of production became necessary with heightened competitive pressure from Asian producers assailing the West and provoking 'lean production', 'downsizing', and such like. Hence the supply chain, while continuing under hierarchical control, as in Japanese *keiretsus*, nevertheless took the vertical network form of out-sourcing but with 'preferred supplier agreements', 'value-added partnering', 'externalized learning' and the like as the paradigmatic form. Simultaneously, another, less vertical, more lateral production paradigm where SMEs predominated, as in northern Italy and particularly in 'new economy' sectors like TMT where SMEs networked intensively in the lateral dimension, clusters became the predominant business model. Most recently, among industries where the possibilities for distance working as freelances from a domestic base are pronounced, most obviously, but by no means exclusively in the software engineering and systems design, consultancy, advertising, new media and cultural industry fields, the 'virtual firm', coalescing temporarily to fulfil a specific project's requirements, has emerged. But project-based working is pervasive through the other three paradigms too. The virtual firm, overhead-free in effect, and run from home with outworking partners domestically and abroad, is an extreme form of a generic, paradigmatic characteristic of the networked economy. But its very fleetingness means the paradigm has shifted back towards arm's-length

	MARKET	NETWORK
HIERARCHICAL	Corporations	Supply chains
DISTRIBUTED	Projects	Clusters

Figure 5.1   Production paradigms – from corporations to projects
Source: Adapted from Mariussen *et al.* (2000).

exchange among competing networks and hence the market side of the equation. Networks were never an alternative to the market, merely a lower transaction-cost variant, the advantages of which have been re-absorbed back into market norms which now take trust and reputation for granted as part of a new and more efficient business model.

From networks to clusters and beyond

Let us start by immediately contrasting networks and clusters, not least because they are often discussed as if they are the same thing, which they are not. From the foregoing discussion it is clear that functioning, actively funded and project-focused networks tend to be relatively small in number of members. This is because firms in networks are either put together for reasons of presumed or demonstrated complementarity as part of a support programme such as those in Norway and Denmark, or they are self-selected for good business reasons by firms, with or without programme support, who wish to work together. Either way, the numbers involved will be limited. Clusters, by contrast, can include 10,000 firms, as in Silicon Valley where over 4,000 were core high technology firms even in 1992 (Saxenian, 1994). In the Baden-Württemberg automotive cluster, as another case in point, there are well over a thousand firms located in proximity as well as in terms of business linkages in the middle Neckar valley and outlying towns beyond greater Stuttgart, the valley's main urban concentration (Heidenreich, 1996). In cases such as these, networks of all kinds exist, but the predominant form of interaction is arm's-length exchange. Where firms are cooperative it is, as Langlois (1993) presents it, yet another form of competitiveness, whereas with networks, firms are rendered competitive collectively in ways impossible were they to remain non-networked. Thus, while networks are likely to work to agreed objectives, clusters are more likely to recognize a shared identity that can be translated into economic value.

These distinctions are captured in Table 5.1 which differentiates clusters from networks, a task conducted earliest by Rosenfeld (1997) but adapted to take account of a wider range of cluster cases than was available at the time. Accompanying work by this author showed that clusters could be envisaged, particularly from a policy perspective, in three ways (Rosenfeld, 1995; 1996):

- Working Clusters are identifiable and function as developed organizational spatial forms characterized by vertical and horizontal network relationships among firms in the same and complementary industry sectors as well as between firms and other business-relevant organizations (e.g. financial, research, training, etc.) in geographic proximity. Such proximate interactions are non-exclusive but of such significance that firms express locational preference for the cluster form.
- Latent Clusters have unrealized potential to become Working Clusters because they have the key features of geographic proximity and inter-sectoral complementarities, but firms and organizations in such settings

have yet to evolve into significantly interacting entities. It is better described as an agglomeration. The agglomeration will continue to enjoy certain important external economies or 'spillovers' from co-location, such as access to a common transportation hub or specific labour market, but these *localization* economies have yet to be transcended.

- Aspiring Clusters are the policy-maker's dream but also nightmare in that actors representing an industry may seek to establish a cluster from scratch or to build one where there are some grounds for optimism. In the worst case, the optimism remains unfulfilled even after sums of development capital have been injected into a cluster-building strategy. In the best case, the policy effort of bringing actors together to overcome a policy perception of arrested development brings benefits from economies of association. Porter (1998) describes a case of the latter in the Boston biomedical devices industry. Technopoles in France and Japan often have the former character because of an absence of 'synergy' (Castells and Hall, 1994).

Thus, we can move towards a strong definition of what constitutes a cluster, based on the differentiation of the concept from that associated with networks, and the characteristics of Working Clusters. Academics can be remarkably reluctant to define clusters. Thus, Steiner (1998) in introducing a collection on the subject, notes some common elements in other unspecified authorial definitions, such as an emphasis on *specialization* around an input–output system or in terms of some asset such as knowledge, an emphasis on *proximity*, and one on *cooperation* between firms and agencies occasioned by specialization and proximity. Another author in the same collection refers to Rosenfeld's work as discussed here but also quotes a definition that limits the defined cluster to a value-adding supply chain (Bergman, 1998). Neither really gets to the heart of the matter and both present a rather static picture. Attempting a dynamic definition, Swann *et al.* (1998) disappoint by defining 'A geographical cluster [as] a collection of companies located in a small geographical area' before defining it in relation to 'a *network firm*, possibly quite small and with little vertical integration' which thrives on the complementary strengths of neighbouring firms by subcontracting It is thus close in intent but not execution to Michael Porter's (1998) definition, which is that: 'A cluster is a geographically proxi-

Table 5.1   Differences between clusters and networks

Clusters	Networks
large scale	small scale, inter-firm
open membership	restricted membership
competitive with cooperation	competitive through cooperation
informal interaction	formal partnership
input–output linkages	interdependence
mainly exchange relations	agreed objectives

Source: Adapted from Rosenfeld (1997)

mate group of interconnected companies and associated institutions in a particular field, linked by commonalities and complementarities.' This is better, but it
is still a remarkably static portrayal when the very feature that makes clusters so
interesting to policy-makers and academics alike is their apparent propensity for
fast growth, high incomes and rapid new firm-formation. Given what was said
previously about the cluster phenomenon being a market, having an identity
and being both vertically and horizontally interactive with firms and agencies,
the following is preferred for completeness. It is a rigorous definition in that it
excludes agglomerations because they lack identity and a capacity for establishing representative associational mechanisms (see Cooke and Morgan, 1998)
but it captures the dynamic element better than the other attempts. Thus, the
preferred definition of a cluster is: 'Geographically proximate firms in vertical
and horizontal relationships involving a localized enterprise support infrastructure with a shared developmental vision for business growth, based on
competition and cooperation in a specific market field.'

In brief, Porter (1998) holds that a number of advantages are derived from
clusters, among these are the following. First, *productivity* gains arise from
access to early use of better quality and lower cost specialized inputs from
components or services suppliers in the cluster. Local sourcing can be cheaper
because of minimal inventory requirements, while transaction costs can generally be lower because of the existence of high trust relations and the importance
of reputation-based trading. Common purchasing can lower costs where
external sourcing is necessary. Serendipitous information trading is more likely
in contexts where formal or informal face-to-face contact is possible.
Complementarities between firms can help joint bidding and scale benefits on
contract tenders, or joint marketing of products and services. Access to public
goods from research or standards bodies located in proximity can be advantageous.

Second, *innovation* gains come from *proximity* between customers and
suppliers where the interaction between the two may lead to innovative specifications and responses. User-led innovation impulses are recognized as crucial to
the innovation process and their discovery has led to a better understanding of
the interactive rather than linear processes of innovation. Proximity to knowledge centres makes the interaction processes concerning design, testing and
prototype development physically easier, especially where much of the necessary
knowledge is partly or wholly tacit rather than codified. Localized benchmarking among firms on organizational as well as product and process
innovation is facilitated in clusters. Qualified personnel are more easily recruited
and are of key importance to knowledge transfer. Informal know-how trading is
easier in clusters than through more distant relationships.

Finally, *new businesses* are more readily formed where better information
about innovative potential and market opportunities are locally available.
Barriers to entry for new firms can be lower because of a clearer perception of
unfulfilled needs, product or service gaps, or anticipated demand. Locally available inputs and skills further reduce barriers to entry. A cluster in itself can be

an important initial market. Familiarity with local public, venture capital or business angel funding sources may speed up the investment process and minimize risk premiums for new start-ups and growing businesses. Clusters attract outside firms and foreign direct investors who perceive benefits from being in a specialized, leading-edge business location. These may also be a further source of corporate spin-off businesses. Hence clustering has become a leading model for economic development affecting, in particular, businesses in the knowledge-based economy but by no means limited to those alone. Indeed in Porter's own cluster analysis of the US regional economy, he identifies some sixty-six in thirty-four cities. These range from carpets in Dalton, Georgia, to sawmills and farm machinery in Boise, Idaho, to golf equipment in Carlsbad, California, and biotechnology in Boston and Silicon Valley. With respect to the first, Krugman (1991) gives an interesting account of the cluster evolution process, beginning with teenager Catherine Evans who, in 1895, made a candlewick bedspread as a wedding present. The tufting handicraft she deployed was all but extinct, but in 1900 in response to rising demand she discovered an innovative way to lock tufts into a backing material. Imitation set in and the technique was popular for chenille garments by the 1920s. The tufting process was mechanized by the 1940s and applied to carpet-making, leading to the formation of numerous carpeting SMEs in Dalton together with specialist suppliers of inputs like dyeing and backing. Today 19,000 workers are employed in all but one of the US twenty top carpet manufacturers in and around Dalton.

Similar stories, one involving another Welsh entrepreneur, John Dagyr, who, according to Krugman, started the Massachusetts shoe industry from his cobbler's shop in 1750, another a Methodist minister in Troy, New York who started the detached collar and cuff cluster in that vicinity, are told about these historic economic formations in many all-but-forgotten accounts. In the Spanish Basque Country, for example, the machine-tool cluster at Elgoibar near San Sebastian, owes its origin to two waves of innovation over a 500-year period. The present 4,000 workers in eighty firms owe their livelihoods to the fifteenth-century founding in nearby Eibar of a royal armaments establishment, a tradition which survived to the nineteenth century when Estarta & Ecanarro began making sewing machines under licence from Singer. There being no local supplier of components, the firm itself made these and the machine tools to produce them, until the presence of a local market stimulated entrepreneurship through vertical disintegration. Estarta & Ecanarro was the direct parent of twelve firms and indirectly through one of these, of a further six. Firms like *Lagun* one of the direct spin-offs from 1969, employed twenty years later some 150 in the production of milling machines, 60 per cent of which were badge-engineered for Los Angeles machine-tool firm *Republic*, the rest being exported worldwide (Cooke *et al.*, 1989).

Of course, Marshall (1916) wrote extensively about the cluster phenomenon, or the industrial districts, as he termed them in his accounts of the evolution of the dominant manufacturing industries of nineteenth-century Britain. Feser (1998) makes the vital point that while Marshall stressed both static and dynamic externalities, it is the latter that are most relevant to our understanding of contemporary cluster formation in 'new economy' sectors because they are the spillovers most closely associated with learning, innovation and enhanced specialization of the kind associated particularly with knowledge-based industries. Thus skilled labour, highly refined input specialization and knowledge transfer are at the heart of today's new clusters though the degree of 'stickiness' of the milieu in which these assets are embedded (Polanyi, 1944; Granovetter, 1973; 1985) also counts. Embeddedness is conventionally understood as a quality present in settings where social capital is strong and, as a consequence certain local synergies are made possible. Small amounts of capital may be raised informally because of the reputation of those seeking it and the strong community ties that reinforce trustworthiness. But Woolcock (1998), following Evans (1995) casts doubt on the appropriateness of embeddedness alone as a developmental mechanism. These authors stress the importance for growth of a correlate to embeddedness they term *autonomy*. This means that to move beyond the start-up phase, firms must integrate with external networks and develop autonomy from strong local ties, not least because of the constrained nature of the support embeddedness alone can offer.

A paradigm case of what might be called weak embeddedness with strong autonomy is related by Biggiero (2000), concerning the biomedical industry in Mirandola, near Bologna, Italy where eighty firms employ 3,000 people. The main business model is to spin-off from an innovator, become a sub-contractor then sell out to a multinational. This practice began with the first-mover, Veronesi, who established his firm in 1963 and sold it to Sandoz in 1973. He did this three times more, selling to Fiat in 1982, US firm Mallinkrodt in 1994, then Fiat again in 1995. Veronesi's protégé, Bellini has done the same three times, selling to the Italian investment firm Ravizza and US firms Braun and Baxter Healthcare between 1978 and 1995. Other entrepreneurs have quickly sold to German, Swedish and Italian multinationals. Embeddedness enables firms to identify medical needs due to close links with local hospitals, autonomy facilitates the financing, production and marketing of products. But the key feature of this cluster is the balance between the two and the serial entrepreneurship that is driven by the cluster's dynamic externalities.

In the UK, research shows there to be some twenty-five working clusters of the kind discussed. These range from Aberdeen's oil industry, employing 50,000 to London's financial cluster employing some 400,000 and to Cambridge's IT cluster with some 4,000 employees. Smaller clusters include opto-electronics in north-east Wales, new media in Brighton and Internet games software in Liverpool, Sheffield and Glasgow (Drennan, 1996; Hendry and Brown, 1998; Cooke *et al.*, 1999; Leadbeater and Oakley, 1999; Tang, 1999). If we look in more depth at some of these, we find that the *project* focus is becoming the predominant 'glue' holding cluster firms in hard networks for a limited period of time, but that such firms will separate, then re-coalesce around new projects, sometimes with new partners added, often with at least a core of familiar partners and sometimes with familiar partners in new roles.

Tang's (1999) account of Brighton's new media cluster of some 110 firms reveals firms to be highly project-focused, highly competitive and not overtly conscious of being part of a cluster as such. This is despite the existence of 'Wired Sussex' a local industry association part-funded by the UK Department of Trade and Industry, and Brighton's Multimedia Development Association (MDA). Several members of the latter have worked on projects such as MEDIALAN, a trial broadband area network, then other Information Society and EU Inter-Regional projects requiring public and private co-financing, to develop new media infrastructures. But most firms are focused on electronic publications, either CD-ROM or online, such as leisure software (games), reference and academic material, business and financial information, image and music libraries and Internet indexes. These are entirely project-based activities based on niche markets. The attractions of the location include a local arts culture, the density of similar firms and the labour market and information spillovers arising from that, and the capabilities of firms to 'feed off' each other in terms of contacts and ideas. The traditional inter-dependence between creative and technical talent is realized more easily in the cluster setting. But Tang, as other researchers often find, had to drag information about social networking for business purposes from entrepreneurs who, as she puts it, 'seemingly perceive these networks as formal-ized activities and exchanges, and do not feel they matter significantly'. In other words, they are simply an established part of the business environment and entrepreneurs think of them, if at all, perhaps as much as they think of their consumption of electricity.

In the Formula 1 motor sport cluster, which extends around London's M25 orbital motorway, from Surrey to Cambridgeshire, two features, short-term, project-based work routines and intra- as well as inter-sectoral learning through embodied knowledge transfer by mechanics and other technical experts working on short-term contracts characterize its success on a global scale. Henry and Pinch (1997) refer to it as benefiting from being a 'community of knowledge' which has caused the foundation in or migration of constructors like Ferrari, McLaren-Mercedes, Williams-BMW, Arrows, Jordan, Ford-Jaguar, Tyrrell and Benetton to the cluster, joining some 115 specialist supplier firms ranging from engine specialists Cosworth Engineering to Guy Croft Tuning and Zephyr Cams Ltd. The swift dissemination of knowledge within the cluster and the application of knowledge from the aerospace industry, which is co-located with part of the cluster geographically, give it competitive advantage. Innovations learned from aerospace techniques include active suspension, aerodynamics, carbon fibre and composites in construction, computerized telemetry, 'fly-by-wire' throttle control and aerodynamic 'spoilers'. Such knowledge flows through the system as engineers, designers and technicians interact socially and technically, but mostly through the churning of the labour market as projects or contracts come to an end or better ones are found elsewhere.

Having said that, the existence of a cluster identity rests more on the efforts of public organizations than the industry itself. Despite Henry and Pinch's (1997; 1999) praise for the 'knowledge community' aspects of the cluster, the reality is that those firms that even knew of each other's proximity, like Williams and Benetton, were at war over poaching and predatory 'learning' through job mobility. It was through local authority economic development department and Training and Enterprise Council (TEC) efforts in Oxfordshire that a representative association, the 'Motor Sport Forum' was established to moderate the conflict and find collective solutions that helped create a meaningful cluster. Thus the Forum was responsible for the establishment of a Masters degree in Motor Sport Engineering at Oxford Brookes University, thus releasing some of the pressure on skills supply. In 2000 it was, with Oxford University, instrumental in the establishment by Oxford Innovation, a technology trust, of an innovation centre for materials science. The university transferred its material sciences department to the facility, which also houses Ford's materials research centre in partnership with the motor sport industry (Willis, 1999). Hence we see in this example an important illustration of the effects of learning by interaction in overcoming weaknesses in arm's-length exchange relations, indeed transforming a knowledge community into a learning economy through creating collective infrastructures for innovation to keep the cluster globally competitive. This is where the advantage of clusters is most evident because they can change the cut-throat competitiveness that can lead to market failure into social capital that, by virtue of collaboration through associational activity, brings competitiveness to new levels.

Conclusion

It was said at the outset of this chapter that an attempt would be made to move from the relatively abstract language of social capital, learning and trust to concrete cases of industry organization that gains advantage from collaborative practices that enhance competitiveness. As Langlois (1993) saw it, there were five varieties of competitiveness, beginning with cost-benefit calculation, substitution of production factors (e.g. capital for labour and knowledge for plant) and non-price competition (such as quality or innovation improvements). Beyond these are customer involvement (after-care service) and collaboration with other firms and organizations. The last of these is now clearly part of the orthodox canon of competitiveness criteria as in the influential work of Porter (1990; 1998) in which collaborative modes of business practice are held to be essential, even superior to stand-alone competition. Networks and clusters are specific modes advocated by business economists, especially where rapid productivity and innovation gains are key features of global competitiveness. These, along with high rates of new firm formation, are pronounced in new economy industries, the so-called technology, media and telecoms (TMT) triad because they are knowledge-driven and often close to the science base. Accordingly, learning is a central feature of the conventions associated with clusters and networks in these spheres. But clusters and networks are by no means limited to the new economy. Most of them are found in the old economy but competitiveness is a generic requirement under conditions of globalization and the cluster approach has been revitalized, due to the need for increased productivity and innovation among sectors that may no longer be dominated by oligopolies as they once were.

In learning economies, where skills must be upgraded and the skills sets of firms changed, it is no longer sufficient to seek to simply buy solutions to short-term problems by poaching workers from nearby firms. As the case of the UK's 'Motor Sport Valley' showed, new forms of governance of industry clusters are necessary to overcome what Ostrom (1992) and others refer to as 'commons' problems, the tragedy of which is that individual utility-maximization destroys community assets. Thus networking and clustering both require a minimum level of shared interest on the part of firms to begin with, but can quickly create commons problems if left unregulated, thus bringing forth new problems which are best addressed with the involvement of third parties. Whereas firms often owe their origins to individualistic entrepreneurship, particularly when barriers to entry are low, if they operate in proximity to one another, gaining 'free-rider' externalities in the process, a point will be reached where they recognize the desirability of developing new externalities. Sometimes these may take the form of public subsidies for re-training or restructuring businesses to face greater competitiveness pressures. This is precisely what happened in one of the world's leading cluster regions in Emilia-Romagna, Italy, when small firms complained to the regional government that they did not have and could not afford computer-aided design and manufacture, something they feared their competitors already took for granted. Innovation centres were set up with the aid of EU funds to help overcome such technology and training problems (Cooke and Morgan, 1998). An advantage in this region was that it is characterized as having high social capital and a trustful disposition on the part of local firms towards each other. In 'Motor Sport Valley' firms started off with a no-trust rather than even a low-trust disposition, but benefited from the trust-building actions of local third parties to develop new positive externalities through economies of association, or what Florence (1958) called 'geographies of association'.

Porter (1998) tells a comparable story about the Boston biomedical industry, one of Massachusetts' statistically more important industries but which was collectively unaware of its own identity, fragmented, individualistic and under-performing. In the 1990s a new administration was elected to run the state, committed to reversing the decline of Route 128 and the tarnishing of the 'Massachusetts Miracle' of previous decades. The cluster idea was a key part of Governor Weld's programme, advised by experts from Harvard Business School, including Porter, and biomedical products was a strong candidate to receive support, not least because of its geographic concentration upon Boston and apparent intra-industry complementarities, strong local and national markets, and skilled labour base. A meeting to form an association was called by the administration's agents and representatives of firms came, but obviously neither knew nor trusted each other to engage in joint actions. No desire for a future meeting was forthcoming; even so, the administration persevered and eventually, after a number of years of meetings, an association was formed, influenced somewhat by the success of the Boston biotechnology cluster, which had benefited greatly from associational activity (Massachusetts Biotechnology Council), not least in persuading the Food and Drug Administration to open an office in Boston, saving significant expenditure on visits previously necessitated to Washington, DC. Whether these third-party efforts were justified remains to be seen and it may be that not all industries sectorally or geographically have high receptivity to clustering. Nevertheless, what can be said with a high degree of certainty is that embeddedness is not a 'natural' feature of co-location and that even when induced, firms may never see any advantage in being part of a cluster, or at least not for a long time.

What this tells us methodologically is that it is clearly no use to define clusters in terms of co-location alone as many studies in fact do. At best, such forms are Latent Clusters, but are better described as agglomerations, in which firms locate because of 'localization economies' like transport or human capital, which they exploit passively rather than seeking to develop as social capital through which embeddedness, interactive learning and innovation may flourish. Policies that sought to co-locate firm functions in the hope that co-location would of itself create cluster synergies have failed, from the Japanese to the French Technopolis programmes to North Carolina's Research Triangle Park (Longhi and Quéré, 1993; Castells and Hall, 1994; Asheim and Cooke, 1998; Longhi, 1999). This is not to say such places have failed to attract highly qualified, high income, high value-adding employment in R&D and related activities into places that previously lacked them, but without localized interactions involving commitment to pursue joint action lines to take advantage of new exploitation opportunities, they are not Working Clusters. This is because they lack localized networks that are the origin of clusters. Without these, they have weak or non-existent means to engage third parties to facilitate the solution of collective problems. Nor can they assist in the establishment of infrastructures to enhance the prospects for innovation by providing collective goods through collaboration but able to be appropriated by individual firms in ways appropriate to their requirements. Clusters thus display characteristics of embeddedness which, through the ties that bind, whether strong or weak, give rise to social capital that is itself the expression of reciprocity, mutuality and return of favours typical of high-trust settings.

But this chapter has shown that embeddedness is not enough to secure significant, notably global, economic development. It was not always necessary historically, though Marshall (1916) showed how central it was in the industrial districts that gave rise to early capitalism, and we are seeing how at the dawning of the Information Age the cluster form seems to be crucial to the evolution of entirely new industries like IT and biotechnology (Swann *et al.*, 1998). Why this form is so central to the survival and growth of such industries is the subject of the following chapter, but certainly learning, knowledge acquisition, particularly of the tacit form which is most easily released and discovered in real-time meetings of the 'talk and listen' variety, and the spillovers that reduce the expense of formalized transaction costs like lawyers' fees for contracts, the so-called 'untraded interdependencies' which are so beneficial to small firms at the early stages of development, are all interwoven in the explanation. In this sense embeddedness and proximity go together to create clusters. However, to develop beyond the cosy confines of the cluster, firms have to develop autonomy by developing external networks, which broaden their functional operating fields as well as their geographical markets. Critics of the embeddedness analysis like Evans (1995) and Woolcock (1998) go further and say that such firms need also to develop *integrity*, valuing membership of professional, non-proximate networks which allow them to learn interactively and become increasingly knowledge-driven through intimate involvement with peers and leaders. To which they add also the need to move beyond even their professional and local peers by building network relations with governmental bodies and agencies, something that, as we have seen, can occur through associational practices in clusters. But *synergy*, as such ties are called by these authors, involves more intimate links with government departments that may be lobbied or accessed to provide support or take soundings about the desirability of regulatory changes. These, of course, may be the hardest network linkages for firms to form, not least because of their time-consuming nature and maybe alien modes of operation.

In the beginning, this chapter posed the question of the sense in which networking and clustering constituted a type of social capital, especially given the discussion in the preceding chapter concerning whether or not it might constitute the missing ingredient in regional and local economic development policy. This also begs a widely asked question as to whether, if clusters and their supporting networks are effective for enhancing productivity, innovation and new firm formation, as Porter (1998) for one asserts; can they be built? The argument developed here and supported with textured case material is that social capital has a strong economic and business development dimension because it emphasizes a previously hidden dimension of competitiveness, namely, collaboration, that firms often find vital to acquiring through learning, new knowledge of an uncodified or only partly codified kind, possibly at zero financial cost, which is a significant asset to business performance. They may also gain scale advantages enabling them to compete in otherwise unattainable markets if they work, albeit temporarily, with partner firms and organizations. Proximity aids this process enormously but it also, through embeddedness, limits horizons, something needing to be guarded against by synergy, integrity and *linkage* to external networks. So it can be concluded that building social capital along the lines presented here is a wise course of action and examples provided underline that judgement.

Clearly, too, networks can be and have been built from scratch through network brokerage programmes of the type described for Scandinavian, North American and other cases. However, three important things have been learned in the process. First, networks work sub-optimally where partners are put together rather than choosing each other. Second, networks produce significant firm benefits when partners pursue concrete projects for which they are fully conscious they need network partners. Third, where policies to support social capital-building through networks are implemented, evaluation of their success or failure should be flexible enough to act on the object of the programme, i.e. whether sustainable partnerships resulted beyond the life of the programme for programme-member firms, rather than be limited to the programme itself, and were these networks directly formed by the programme? For clusters, however, the evidence that they can be built from scratch through policy action is less evident. Clearly, growth poles and technopoles have been built, but these are not clusters. Equally clearly, clusters of specialized, learning and interacting firms have come into existence from zero as a consequence of entrepreneurial activity, but scarcely any of these have been built by active cluster policy. Hence, clusters can be and are commonly built, probably but not incontrovertibly with some informal or maybe even formal networking entailed, but not often led by policy. Policy can help when some concrete assets and opportunities for development are observed, probably by the policy community itself, rather than firms in the Latent Cluster. But examples of policy-led cluster formation are hard to find. However, in the chapter which follows some of the few instances of this that have been identified, along with more conventional accounts of those that have arisen through entrepreneurship, then been enhanced with public and other third-party involvement will be presented and some evidence adduced that cluster-building from near to ground zero is not impossible.

6 The clustering phenomenon in the 'New Economy'

An anatomy of growth and further lessons for policy

Introduction

In the previous chapter the distinguishing features of business clusters were defined and contrasted with those of business networks. In this chapter space will be devoted to examining why clusters are such a prominent organizational mode for firms engaged in what is called the 'new economy'. This terminology is itself new and the next section is devoted to an explanation of some of its key features. It is important to do this and to examine the kinds of business active in new economy clusters since among these are likely to be many of the growth firms and industries of the future. Whether firms or clusters of new economy firms will survive as such for long or eventually be absorbed through merger and acquisition activity into old economy sectors is a matter for debate. Some, especially dot.com businesses, seem vulnerable because their business model is based on revenues from advertising and a customer base, both of which can disappear very swiftly. They are doubly vulnerable when highly valued on the stock market but have yet to earn profits, because investor confidence can evaporate more rapidly than a customer base. However, this model, whereby investors are persuaded to invest large sums on the expectation of future profits is one that first started with biotechnology firms. Some analysts think they have under-performed over the twenty-year period they have been in operation, but despite peaks and troughs in investor confidence, stock valuations and investor interest remain generally high. In areas where recent efforts to induce growth by virtue of government subsidy have occurred, Germany being the most obvious case, investment and venture capital have flowed into the subsidized sectors, especially biotechnology.

A substantial part of this chapter explores that process, the prospects for success and the future of policies to induce clusters. This is important because the theoretical burden of Chapter 1 of this book is that capitalism is a profoundly disequilibriated form of economic organization, and clusters are among its presently most highly developed, new forms. Much more than in the case of the earliest rounds of development, the new economy clusters are highly skewed in geographical terms. They are overwhelmingly found in or near large, well-diversified services and knowledge-based cities or specialist research university campus cities or towns. In contrast, these knowledge-driven clusters are not found in rural locations, nor are they found in older, declining industrial locations of the earliest phases of the 'Industrial Age', nor are they found in the still quite extensive urban and regional systems founded on the later 'Industrial Age' of Fordist consumer industries such as automotives and consumer appliances. This is despite the fact that some of these 'old economy' industries will display residual or even renewed cluster-like features, dating from their foundation or more recent restructuring. But these are mature clusters with few strong links to 'new economy' clusters like biotechnology and other telecoms, media and technology (TMT) sectors, though this is not ruled out in future. Part of the emergence of new regionalism in recent decades is conditioned on a perception that old centralist remedies have run their course and are, by the twenty-first century, redundant. A belief that decentralized industrial policy is more likely to bear fruit is widespread and represented in the growth of regional economic governance in Europe especially, catching up with parts of the world with newer founding constitutions based on federalist systems. In this chapter an effort will be made to explore whether new regionalism does or can have a serious effect on the processes underpinning the often spectacular growth of knowledge-based industry in the new economy or whether some other kind of economic development based on continuing regional economic dependence lies in store. But, first, attention is devoted to new thinking about the new economy and aspects of its cluster support infrastructures.

What's new about the new economy?

One of the first extended discussions of the nature of the new economy was presented in a book by Kelly (1998). Essentially a book about networking and the likely impact of the Internet on business, it concludes with ten rules for the new economy that have wider application. These are implicitly presented in contrast to the rules of the old economy and Table 6.1 presents them in that fashion for comparison with a further set of new economy 'conventions' (after Kaplan, 1999) to be presented below. In Kelly's itemization of the rules in

Table 6.1 New rules for a New Economy

Old Economy	New Economy
centralized	decentralized
constant returns	increasing returns
value scarcity	value abundance
rising prices	falling prices
maximize firm value	maximize network value
incremental innovation	disruptive innovation
place proximity	cyberspace
machine-focused technology	human-focused technology

Source: Adapted from Kelly (1998)

question, repetition means the ten, in fact, become eight. They relate to the main organizing features or assumptions about rational business practices and expectations that arise from their pursuit. Rather typically they are stated in binary opposition to presumed rules operating in the old economy and this is no doubt overdone. However, if they are used as emphases rather than route-maps they can have, at least, some heuristic value. First, technical networks and the human ones they facilitate, overcome the centralization implicit in classic corporate and governmental bureaucracies. In principle this has some validity but it is patently clear that the power that has 'swarmed away from the centre' has often been less commanding executive power and more marginal, individual influence. In a few cases and specific industries, traditionally powerful corporations have been supplanted by new, rapidly growing companies, most obviously in IT and the toppling of IBM by Intel and Microsoft. But the decentralized power of the latter two is mainly over technology not financial rectitude, as the notorious 'control freakery' of Grove and Gates make clear (Jackson, 1998; Edstrom and Eller, 1998). Kelly says the new economy overcomes scarcity and yields increasing returns to scale. The latter now has a respectable pedigree following the discoveries of the 'new neoclassicals' led by Krugman (1991; 1995; for comment, see Cooke and Morgan, 1998) and clusters are said to be the means of inducing them to some extent. But clusters thrive on scarcity of intellectual capital, tacit knowledge exchange and the immobility of such assets. Hence the new economy continues to be based on scarcity, but of knowledge 'capital' more than, for example, financial capital.

So prices of computers and software fall in real terms but not drugs, especially not biotechnologically derived drugs and, overall, modestly rising inflation remains an accompaniment to the new economy overall. However, to be fair, technology-influenced productivity increases have begun to be apparent in some areas of the new economy and parts of the old one integrating with it, such as telecom services, logistics and, in pure cost per item terms, e-commerce. Incremental innovation remains overwhelmingly predominant over disruptive or radical innovation even in knowledge-based industries, making the new economy more disappointing as a source of truly protean innovation than the Industrial Age, according to Krugman (1995). Networks, as we have seen, are an important instrument in competitiveness, and add value under some but by no means all circumstances. Place proximity is important for clusters, but if the analysis in the previous chapter that foresees project-based work as likely to supersede them is correct, the question is, does cyberspace substitute? For reasons already discussed, it is codified knowledge that transcends space easily but not new knowledge that is often created by communities of distinctively skilled people exploiting spillovers in specific knowledge-intensive places. So, as with much of his analysis, particularly the assertion that technology becomes anthropomorphic in the new economy (think of call-centres), Kelly has overstated his case but not without pointing to some possibly important tendencies present in TMT industries.

A more sober analysis is offered by Norton (2000) who makes the important, if fairly obvious point that the new economy is remarkably geographically focused, is highly Schumpeterian in its particular occurrences, and that it shows extremely strong tendencies towards disequilibrium. Interestingly, he has investigated why this should be so and found by careful statistical analysis the cluster-focused disequilibria that characterize the spatial incidence of the new economy. It is, argues Norton, a real phenomenon and its propulsive power was networked IT, the increasing returns from which are likely to widen the IT lead between the USA and the rest of the world. This judgement would have to be modified were it to move beyond networked personal computers (PCs) to mobile telephony since, technologically and in terms of market penetration the US lags behind several EU Member States. Like Kelly, he sees the decline of centralization and hierarchy as accompaniments of the new economy, stresses the importance of free-flowing information, though does not comment on the barriers to that identified as 'asset stock accumulation' and willingness to trade information only with others of value equivalence (see Malmberg and Maskell, 1997). He also explores the importance of geographical proximity for the entrepreneurial innovation that is hypothesized to lie at the heart of the new economy. This is where he offers an anatomy of the continuing success of Silicon Valley as an innovative cluster, concluding that tolerance (of failure and treachery), risk-seeking, restlessness, reinvestment in the cluster, meritocracy, collaboration, variety, product-obsession and low entry barriers comprise the culture of this economic community, the capital of the new economy, 'a milieu conducive to spin-offs and start-ups' (Norton, 2000, p. 239). His conclusion is that Silicon Valley and other, lesser, though also new economy places are characterized by the geographical concatenation of scientists, engineers, entrepreneurs and venture capitalists looking for value from technological discontinuities, the more disruptive, hence rarer, the better. In particular, places with high digital IPOs (initial public offerings) like San Francisco, San Jose, Denver, Boston and Seattle have high concentrations of the key new economy actors.

This is interesting because it links to the third cut we will take at tackling the new economy and trying to assess its ephemerality or otherwise. In Kaplan's account of the rise of firms like Cisco Systems, Netscape, Yahoo and Oracle, all Silicon Valley-based, great emphasis is placed on the likes of Kleiner Perkins, Caufield and Byers and other, lesser venture capital houses such as Sequoia Capital, Sierra Ventures, Technology Venture Investments, New Enterprise Associates and the Mayfield Fund, all clustered in Sand Hill Road, Palo Alto. Kleiner Perkins (KP) investments (some 230 firms in which equity is retained) have market capitalizations worth $125 billion, 1997 revenues of $61 billion and employ 162,000 people, mostly but not exclusively in Silicon Valley. In brief, the key role in entrepreneurial innovation is now taken by an aggressive scouring of research laboratories by venture capitalists (VCs), some of whom, like KP have, in effect, built their own clusters of start-up firms who are encouraged (Bronson, 1999, says 'coerced') to trade with each other in Japanese *keiretsu* style. This phenomenon is more advanced in IT, telephony, software and dot.com fields than biotechnology, but since these are co-located and co-funded by the same VCs in Silicon Valley, the model is distinguished mainly by the higher burn rates and slower IPO progress of the latter. Essentially, if a biotechnology business opportunity looks like a winner, the experience is indistinguishable. It is also less the case in biotechnology that VCs are the supplicants, pressing their claims to be allowed to invest, a reversal pioneered by Jim Clark of Netscape, notably with his new Healtheon venture to develop online systems to revolutionize American healthcare (Lewis, 2000).

Key elements of new economy conventions are shown in Table 6.2, borrowed by Kaplan (1999) from KP's slide-show. Many do not differ significantly from our earlier comparisons, but some elements focus emphasis on peculiarities of business practice, like low litigation, intelligence (distributed), and the widespread use of stock options as compensation. As we have seen, the idea of a 'new economy' can be criticized for overstatements of singular features of some industries as being generic, and flimsy evidence that the knowledge-intensive sectors involved have conquered scarcity and are thus immune from the business cycle. However, there are new features to the TMT innovation system, principally the voracity and abundance of investment capital, the rationale for expenditure of which rests on calculated risks that can appear and indeed turn out to be massive and misplaced gambles. To some extent the long lead-times and high cash 'burn-rate' in biotechnology triggered this kind of high up-front cost investment based on a calculated risk that enough returns would accrue when some firms reached the IPO stage to compensate for the losers. Thus innovation is the fundamental source of value, seeking it out is an investment imperative, and systemic search and selection procedures by VCs are the main means of exploiting gains from public investment in basic research. This is the fundamental feature of the new economy, rather than the conquering of scarcity. It is, in fact, investment based on the apotheosis of scarcity, the 'breakthrough' innovation, the 'magic bullet' cure and the chance to reap riches that yield lifetime security. Knowledge-driven TMT clusters help make this happen. However, inspection of the Table 6.2 begs the question in general of how many of the conventions exclusively assigned to the new economy actually apply in the old, whether currently or, more tellingly, when they were new. Thus a strong emphasis on teamwork, multiskilling, customization, agility and distributed intelligence could be said to be characteristic of many mature sectors responding to global competitiveness challenges, just as growth and risk-taking were pronounced features of the early automotive or aviation industries. The emphasis on stock options for workers and its apparently higher ranking than wages is novel, and if true, the rejection of litigation in favour of investing in the newest new thing (Lewis, 2000) is too, although Microsoft must be one of the exceptions that proves the rule. But having made these points, there are some points of agreement among all three authors whose work on the new economy has been compared and contrasted.

First, the idea of decentralization of control of parts of the new economy away from the corporate behemoths of the preceding generation is a fundamental point of agreement. That is, in the newer sectors, based as they are on

Table 6.2   Old and New Economy conventions

Old Economy	New Economy
a skill	lifelong learning
IR conflicts	teams
environmental conflicts	growth
security	risk-taking
monopolies	competition
plants	intelligence
standardization	customer choice
litigation	investment
status quo	agility
hierarchical	distributed
wages	ownership/options

Source: After Kaplan (1999)

clusters around universities, in the main, there is less dependence for knowledge-exploitation and innovation upon the corporate R&D laboratories from whence most commercial innovations continue to be forthcoming, but less so in new economy sectors. Second, the idea of flows of value, particularly of knowledge or innovation capability, through networks, is fundamental. Indeed, so much is this a feature of new economy clustering that in the exemplar of Silicon Valley, the strategy of building 'private clusters', corporate *keiretsus* or EcoNets as they are also known, is not confined to KP. Intel has a powerful corporate venturing arm, Intel Capital, as do Lucent Technologies, AT&T Ventures and Cisco Systems, the latter internalizing its EcoNet by acquisition. Table 6.3 gives an indication of the main corporate venturing actors by number of IPOs in 1999 with the 'if-held' value of the equity involved over the year. These and many other venture houses are key drivers of the new economy, and increasingly the structuring of clusters. They are central with the technology entrepreneurs in the creation of a new kind of systemic innovation that is the real source of advantage over, for example, European competitors, something to which we return at the end of this section. Finally, there is some measure of agreement about the distinctiveness of innovation around new knowledge, moving rapidly from workbench to generating investment if not sales in the new economy. Such is the power of innovation where serial innovators like Jim Clark of Netscape are concerned that they audition venture capitalists for their next IPO, not the other way round. This is the central feature of the new economy as it has emerged thus far. It is not well developed in Europe, though we shall see that developments of cluster-based, venture capital-backed new economy activities are beginning to emerge. But if the European Commission, concerned at the innovation gap with the USA (CEC, 1995; 1997) seeks an explanation for its slow rate of commercialization of science and technology, it needs only to consider the abundance and proactivity of the innovation support system in California or Massachusetts compared to that nearer home.

This brings us conveniently to a brief consideration of the New Economy

Table 6.3 Number of IPOs and value, top ten, 1999

Corporate Venturer/Venture Capitalist	IPOs	Value ($b)
Access Technology Partners	28	73
Intel Capital	23	52
New Enterprise Associates	20	66
Kleiner Perkins	18	78
Comdisco Ventures	17	29
Benchmark Capital	15	113
Goldman Sachs	12	29
Technology Crossover Ventures	12	26
Institutional Venture Partners	11	91

Source: Craven (2000)

Innovation System (NEIS) discussed so far, in relation to the kind of regionalized innovation systems (RIS) that emerged in support of old economy regions, often confronting economic crisis, and have been the subject of much academic and policy interest of late (see, for example Braczyk *et al.*, 1998; de la Mothe and Paquet, 1998; Acs, 2000). Table 6.4 seeks to capture aspects of these distinctions. The key differences are the often public nature of the typical regional innovation system with its technology transfer bodies, science parks, partnership funding, linkage of innovation to hierarchical supply chain relations and strong user-driven, incremental emphasis. This is typical of systemic innovation in Europe (Cooke *et al.*, 2000) and in old economy regions of the USA (Shapira, 1998). The key differences lie in the attitudes of financier and entrepreneur in new economy settings. The cluster is there to be scoured for innovative ideas and potential businesses from the investor viewpoint and there to enable the innovation entrepreneur to accumulate very large sums of money from his or her point of view. Technology is a means to the latter end, though employee stock-option holders may be product obsessives as Norton (2000) suggests (on this, see Bronson, 1999). Whatever assists this process, and incubators staffed with managers who can take the pain out of management for technology entrepreneurs are a case in point, is provided as far as possible.

Hence, an assessment and variable perspective on the new economy have been presented. The interim conclusion is that a distinctive mode of induced innovation has been established and that it has proved to be effective at raising the rate of new firm formation around the new economy or TMT sectors, but really, thus far, only those sectors. This has given the USA, where the model was set in place in Silicon Valley in the 1970s but developed significantly and bravely, some would say recklessly, in the 1990s, a lead which may yet expand further in the first decade of the twenty-first century, except in mobile, convergent (multimedia) telephony where Europe leads. The NEIS is strongly clustered in virtually every case that has been recounted and even if the cluster is not yet properly formed, firms still agglomerate around universities or centres of creative knowledge like film studios. Learning is, of course, the central attraction

Table 6.4 Aspects of regional and New Economy innovation systems

Regional Innovation System (RIS)	*New Economy Innovation System (NEIS)*
research & development-driven	venture capital driven
user-producer relations	serial start-ups
technology-focused	market-focused
incremental innovation	incremental and disruptive
bank borrowing	initial public offerings
external supply-chain networks	internal EcoNets
science park	incubators

where knowledge capital can have rapidly escalating value. The more knowledge-based clusters, such as those explored previously and in what follows, thrive, the more disequilibriated the economy is likely to become spatially and in distributional terms and the more important it becomes to seek ways of moderating this without killing the golden goose. This is an important challenge confronting economic policy-makers everywhere for the foreseeable future. We now turn to an analysis of three distinctive types of New Economy cluster.

New economy clusters: military market-led

The model for this kind of cluster is Silicon Valley, which most accounts suggest could not have begun without large American public sector defence expenditure (Hall and Markusen, 1985; Saxenian, 1994), but the specific example selected is 'Telecom Corridor' in Richardson, Texas. This relatively unknown cluster in the Dallas–Fort Worth Metropolitan Area has a highly concentrated group of some 600 firms employing some 40,000 in Information and Communication Technology, of which over half are in telecoms manufacturing and services. Some of these are among the largest such firms in the field but they operate in a strong milieu of telecoms SMEs. There was relatively little of the associational support normally found in functioning clusters until 1994 and the establishment of the Technology Business Council (TBC) run by the Chamber of Commerce as a non-profit association of ICT firms. It is a typical, specialist, business-based lobbying and networking organization described as a prime example of 'co-opetition' by Richardson Chamber of Commerce (2000). Thus the board of directors has representation from executives of local firms that compete for customers, employees and innovations but cooperate on TBC. It has 300 members defined as Technology Members and Provider Members, with a limit of 20 per cent on the latter. Technology Members include those who research, patent, manufacture or manage technologies, technology integrators, and consulting engineers. Provider Members are service providers to the cluster such as patent lawyers, universities and venture capitalists with an ICT focus (see Figure 6.1).

As in the case of the UK's Motor Sport Valley in Oxfordshire, one of the first acts of the forum was to establish a Technology Training Network (TTN). This links two local Community Colleges with TBC to develop two-year associate degree programmes in technician training for ICT firms in the Telecom

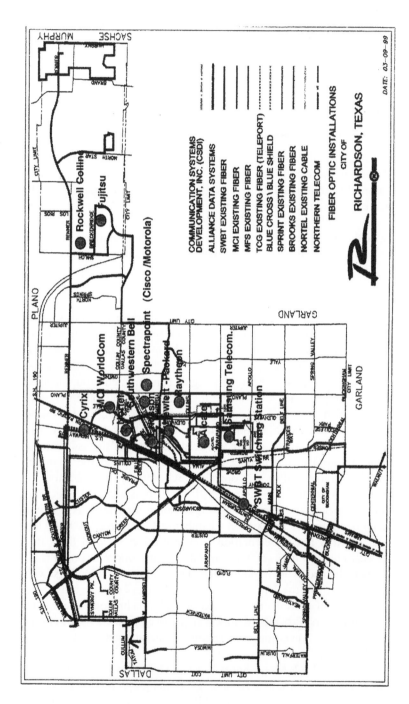

Figure 6.1 Telecommunications cluster in Richardson, Texas

Source: Richardson Chamber of Commerce, prepared by Vanessa Weigelt

Corridor. The state of Texas made its largest ever grant of $1.9 million from its Workforce Skills Development Fund to support training of 1,800 ICT technicians 1996–8. Subsequently, leading firms in the cluster each committed $100,000 to form a consortium with the University of Texas, Austin, University of Texas, Dallas, Texas A&M University and Texas Tech University to train telecom engineers and researchers. By late 1998 an e-jobs website had been established serving some 20,000 telecoms job-seekers and 121 firms. The emphasis placed on learning and skills requirements of the cluster shows clearly how the mere co-location of large numbers of firms in the same or related sectors creates 'commons' problems that are difficult if not impossible for firms to solve individually without entering a compensation bidding war. It is also instructive that technician training was seen as the key bottleneck to overcome and that recourse was necessary to public subsidy to overcome market failure, even in Texas.

The reputation of a cluster is often headlined in terms of the well-known companies operating there, although the backbone is often in the unsung SMEs that spin out or start up often as suppliers to the larger customer firms. Of course the large enterprises are also customers and suppliers to each other within and beyond the cluster as well as sometimes collaborators as we have seen. Table 6.5 lists the largest ICT employers in the Richardson cluster.

The last piece of the cluster analysis for the market-led Richardson Telecom Corridor is to offer an explanation of its origins and growth dynamics, tracing linkage of an originating or on-going nature between cluster members at the firm level. Two companies were key to the early evolution of the Telecom Corridor, Texas Instruments and Collins Radio. The former owed its origins to the oil industry as an equipment and services supplier, hence its Dallas location, but of course it became best-known as a warfare technology firm in the 1940s, headquartered in Richardson, and thereafter the employer of Jack Kilby who invented the integrated circuit. Texas Instruments (TI) grew to be one of the world's early leaders in the semiconductor design, manufacturing and

Table 6.5 Major ICT employers in Richardson, Texas

Company	Employment	Business line
Texas Instruments	9,000	digital signal processors
Nortel	8,000	digital telecom networks
MCI WorldCom	3,800	telephone services engineering
Ericsson	2,900	telecom systems
Alcatel	2,800	telecom equipment
Fujitsu	2,000	telecom equipment
Raytheon	900	defence systems
Rockwell Collins	750	communications and avionics
Hewlett-Packard	650	supercomputers
Samsung	350	wireless communications

Source: Richardson Chamber of Commerce, 2000

applications business. The firm was influential in founding the Dallas site of the University of Texas in 1961.

Collins Radio was a microwave transmissions firm that also became a defence contractor, as a consequence of which it was dispersed by federal policy in the Cold War from Cedar Rapids, Iowa to Richardson. By 1969 it was also supplying lunar television transmission equipment to NASA. Collins was acquired by Rockwell in 1972. At the time, downsizing at Collins and opportunism at TI saw a notable rise in technology start-up businesses, many housed in former Collins facilities. Nortel arrived by acquiring one of its telecom switch suppliers, Danray, also one of the TI start-ups. MCI moved to Richardson for proximity to one of its key suppliers, Rockwell. Further spin-off and start-up activity ensued with opportunities for telecoms market entry following the break-up of AT&T in 1984. This led to further acquisition of start-ups and independent foreign direct investment from the likes of Ericsson, seeking a US headquarters, and Fujitsu seeking proximity to its customer, MCI whose 1980s' growth attracted numerous other suppliers. Alcatel entered the scene in 1991 by acquiring Rockwell's civilian telecoms business. Convergence between start-ups from TI and Rockwell led to new hybrid technology firms and the annual start-up rate was 20–40 in the 1990s. The imperative for suppliers to be in proximity to customers has been an important driver of cluster growth in Richardson during its more civilian phase of technological growth and development.

Underpinning all this as well as stimulated by it, the hard infrastructure for telecommunications is one of the most sophisticated in the USA (Malecki, 2000). There are some 400 miles of optical fibre and Southwestern Bell has advanced switching and a Company Network Operations Command Centre (CNOC) which gives 24-hour, manned monitoring of telephone traffic. There are over 180 Synchronous Optical Network (SONET) rings, with sophisticated self-healing capabilities in cases of technical breakdown. All in all, Richardson has a cornucopia of technological advantages, though the one that gives it a unique edge as an attractor for telecommunications businesses, according to Malecki (2000), is the presence of a packet gateway giving direct access to the Internet backbone. This is rare in the USA as is MCI's point of presence (POP) at the same street junction. This combination of capabilities makes Richardson a key node with the highest-level connections to the global telecommunications infrastructure.

Locations such as Richardson attract TMT businesses because of the quality of infrastructure, thus these assets become a resource that is itself scarce and consequently an important location factor. This double-feedback form of causality is essentially Schumpeterian in that an innovation occurs and causes swarming by imitators. But the ensuing critical mass gives lobbying power to the incumbents if exercised politically in an associative manner. Thus a governance structure, albeit light in weight in this specific, highly entrepreneurial setting, is a vital part of a cluster's capability to embark on an upward growth trajectory, even after the reduction of military expenditure budgets and loss of policies that may once have been of considerable influence on the early growth phase of the cluster. This gives a strong hint that if clusters are hard to design into existence, they may come into existence as an unanticipated consequence of government intervention at a crucial phase. Moreover, in the context of Richardson, the deregulatory impact of the anti-trust case against AT&T also created substantial market opportunities by lowering barriers to entry for spin-off and start-up enterprises. Thereafter, the issue of proximity climbed up the agenda for both larger and smaller businesses seeking to be close to customers and supplier to capture spillovers and other externalities such as untraded interdependencies, tacit knowledge exchange and lowered transaction costs. This is a smaller-scale, technologically distinctive version of events that drove the evolutionary process from military to civilian markets in Silicon Valley.

That military expenditure focused in a particular space at a particular time need not necessarily lead to the evolution of a cluster such as Silicon Valley or Telecom Corridor is testified to by the case of the M4 Corridor in South-East England. Here, as in California and New England at least 40 per cent of UK government military procurement expenditure has traditionally been spent. In Castells and Hall's (1994) analysis, this created agglomeration economies based on the presence of skilled labour and sub-contracting services that attracted foreign investment, especially from the USA. But unlike Silicon Valley, lock-in to easy defence contracts became pervasive and innovation was stillborn. One of the few exceptions to this was the defence contractor in wireless radio, Racal. This firm occupied a company-town setting in Newbury, but through alliances with Ericsson and others created the European mobile telephony infrastructure via spin-off firm Orbitel (Cooke *et al.*, 1992). Thereafter the civilian side of Racal's business was also spun off to become Vodafone, in early 2000 the world's fourth largest firm by market capitalization following acquisition of AirTouch from the USA and Mannesmann from Germany. But Vodafone operates a dominated labour market strategy and, until gaining planning permission to build a headquarters facility in the Berkshire greenbelt, occupied some fifty separate buildings in small-town Newbury (Gordon and Lawton Smith, 1998). The aforementioned authors allude to the absence of an ICT cluster in the Thames Valley, which is the epicentre of the M4 Corridor and draw attention to the way inward investment was encouraged by building industrial parks within a tight labour market setting in close proximity to Heathrow Airport.

Accordingly, global firms in specialist software engineering have established UK and European services centres. A case in point is Oracle, a firm that specializes in corporate software, particularly Enterprise Resource Planning systems. Oracle now employs 3,500 people on an industrial park near Reading railway station. Other ERP firms are also found next door to Oracle, including PeopleSoft, J.D. Edwards, Baan and in nearby Basingstoke, the German firm SAP. Thus the five leading ERP firms in the world occupy adjacent service locations. But they interact very little or not at all and source mainly low-value services locally. Research into this feature of the agglomeration showed that, to the extent they communicated collectively it would be to express concern individually to an economy-wide organization such as the Thames Valley Economic Partnership (TVEP) about labour shortages, and Microsoft, also next door to Oracle, might discuss with them an optimal location of a railway station on the new, sustainably-designed Green Park in Reading, where it is hoped they might expand. But anything beyond these superficialities would be off-limits. However, one effect of the skills shortage for software technicians articulated more widely to TVEP was to stimulate Newbury College to put together a £27 million sponsorship arrangement led by Mentor Graphics, to set up Inpaq, a unique software engineering course for advanced printed circuit board (PCB) design. Nearby Reading College repeated the recipe in gaining sponsorship from Hewlett-Packard for a new computing technical degree course. In conclusion, these actions indicate the individualistic approach typical of problem-solving in the M4 Corridor (Cooke *et al.*, 1999b). In consequence, while the area gives the appearance of operating as a cluster, it does not in fact do so. The defence nexus alone, therefore, was not sufficient to bump-start a cluster development process with the virtuous circle of endogenous growth seen in Silicon Valley or Richardson because of the absence of the innovative entrepreneurship relation between ICT engineers and venture capitalists found in those locations.

New economy clusters: civilian market-led

The M4 Corridor story is interesting and instructive because it points to the limits of proximity, lessons that were also learned in France and Japan where technopoles did not produce clusters. It is mainly an agglomeration of disconnected corporate headquarters and services centres interspersed with public sector research establishments that happen to be in technologically advanced fields, many in ICT, but who see or have little reason to collaborate with each other because the knowledge base and learning economies they rely upon are mostly occurring somewhere else within the corporate hierarchy. The public sector research establishments have had no incentive to engage in commercialization and start-up activity, though this may change in future, and even though some have been turned into more commercial privatized agencies, they are often not as advanced in research terms as the university sector. Finally, within the M4 Corridor there is only one reasonably long-established university, at Reading and it has few if any research linkages with the agglomeration.

However, not so far away the story is quite different and some of the most productive research-led cluster formation is under way. The two cases to be explored here involve biotechnology, one in Cambridge, Massachusetts, the other in Cambridge, England. In Cambridge, Massachusetts, and beyond the resurgent Route 128 as far as Worcester and the I-495, biotechnology businesses can be observed evolving almost in cluster life-cycle terms. From small beginnings in rented, sub-let space on MIT science park or some former industrial or newer office building, twenty or so new firms, often launched with the backing of MIT Technology Liaison Office support, form each year. But clusters, as we have seen, are more than firms and buildings. The science base is exceptionally strong in the Massachusetts Institute of Technology (MIT), Harvard University, Boston University and Massachusetts General Hospital. Each year some $770 million in basic research funding flows through the system. Leading scientists and academic entrepreneurs, one of whom has been involved with some 350 patent applications, are present. Massachusetts has at least 150 venture capitalists, most of them in Boston or Cambridge. There are 132 biotechnology firms in the Greater Boston area (59 in Cambridge), 86 outside, employing 17,000 people in total. Finally, there are numerous intermediary bodies supporting the industry at state level, one of which, the Massachusetts Biotechnology Council is an industry association which organizes common purchasing and other services such as promotion, educational placement and careers development for its 215 member firms. The geographical breakdown, bearing in mind the 59 firms in Cambridge is as follows: 132 firms are located east of Route 128 (59 in Cambridge, 16 in Boston, the remainder between there and Route 128), 58 are located between Route 128 and Route 495 (including 11 in Bedford and 6 in Wilmington) and 25 are located west of Route 495 (including 11 in Worcester). Many of these, especially in the outer locations, are based on science or technology parks, as are many start-ups on the technology park campuses of the key universities.

The market segment breakdown is that 34 per cent of firms are in the therapeutic products sector (meaning they have grown beyond the early stages, typically in platform technologies, including diagnostics), 20 per cent are in scientific equipment or supplies; 15 per cent are in scientific services; 14 per cent in human diagnostics; 10 per cent are in environmental and veterinary; and 7 per cent are in agricultural biotechnology (animal, plant, diagnostic and transgenics). Perceived industry growth areas are in medical therapeutics (genetically produced protein, vaccines, gene therapy, human growth hormones); human diagnostics (monoclonal antibodies, biological imaging, DNA probes, biosensors and polymerase chain reaction); agro-biotechnology (nutraceuticals, rapid diagnostic testing and transgenics) and BioInformatics (biological discovery, patient databases etc.). Seventy-nine firms were founded in the 1980s including (> 300 employees) Biogen, Genetics Institute and Genzyme. A further eighty-eight began between 1990 and 1997, the remainder are more recent start-ups or inward investments. Employment grew from 7,682 in 1991 to 16,872 in 1998. As the industry matures, the number of start-ups is decreasing annually. Between 1996 and 1999 seven mergers and acquisitions occurred. Financing of companies in biotechnology is high risk and analyses show that public investment is present at the risky process or product development stage, often in Federal Small Business Innovation Research grants, competitively bid for and assessed by the National Science Foundation.

Some key elements in the innovation system in the Cambridge biotechnology sector would also include the following public and private enterprises:

- Massachusetts Economic Development department, responsible for R&D tax and capital investment credits, provider of seed-corn subsidies to start-ups.
- MIT, for leading edge biotechnology research and commercialization, through campus incubators, technology park, MIT Entrepreneurship Centre, MIT Technology Licensing Office, links-trained entrepreneurs, through patenting to VCs.
- Harvard University; numerous biosciences, genetics and medical research and graduate training programmes.
- Massachusetts General Hospital and Boston University: research and commercialization at Boston University, Bio Square Technology Park.
- Whitehead Institute of Biomedical Research: an independent research and teaching institution (affiliated to MIT in teaching). World leading research in genetics and molecular biology. International co-leader in the Human Genome project, source of comprehensive, published genome data; technology licensing programme and start-up scheme.
- Massachusetts Technology Collaborative: state-founded, independent body to foster technology-intensive enterprises. Cluster-building strategies.
- Massachusetts Biotechnology Council: trade association representing biotechnology firms (162 full and 53 associate members), provides educational, careers, and promotional information to the industry and conducts common-purchasing contracting for biotechnology firm members.

In conclusion, leading *exploitation* firms such as Genzyme, patenter and inventor of the therapeutic product that controls the genetically caused Gaucher's disease, are closely intertwined with this generation and diffusion system. Moreover, Genzyme as a founder member of the Partners Healthcare System with Brigham and Women's, and Mass General hospitals on research funded at $400 million by the National Institutes of Health, reinforces the system. Along with Biogen and Genetics Institute, plus other internationally known firms such as BASF, Corning and Quintiles and a host of SMEs and start-ups, this means the Greater Boston region is supported by the generation and diffusion organizations and associations already noted, and clearly functions as a well-integrated regional innovation system based on a cluster of leading-edge biotechnology businesses.

In Cambridge, UK, by contrast, starting its Science Park twenty years after Frederick Terman started Stanford's, there are twenty-five biotechnology firms in and around it, including neighbouring St John's Innovation Centre. As in Boston, but a scale of magnitude smaller, firms are growing on a number of new science and technology parks to the south of the university city down the main highway links to London and its Stansted and Heathrow Airports. There are also comparable private, in the main, and public enterprise and innovation support systems making Cambridge (like Oxford) relatively private New Economy Innovation Systems. In 1998 there were some 37,000 high technology jobs in the area comprising 11 per cent of the Cambridgeshire labour

Table 6.6 Shares of biotechnology and services function

Biotechnology firm distribution (%)	Biotechnology services distribution (%)
biopharmaceuticals 41	sales and marketing 29
instrumentation 20	management consulting 23
agro-food bio 17	corporate accounting 15
diagnostics 11	venture capital 15
reagents/chemicals 7	legal and patents 8
energy 4	business incubation 10

Source: ERBI (1998)

market. South Cambridgeshire had about 66 per cent of these jobs while Cambridge city accounted for most of the remainder. The main high-tech activity is R&D, supplying 24 per cent of total high-tech employment, electronics has 17 per cent, computer services, 13 per cent, scientific instrumentation, 8 per cent, and next comes biotechnology, fifth in line at 7 per cent. Probably an estimate of some 2,600 employees in biotechnology for the county is a not unreasonable figure. This would give a Cambridgeshire figure of around fifty core biotechnology firms. The growth in number of biopharmaceutical firms has been from one to twenty-three over the period 1984–97, an average of just under two per year, but the rate was more like four per year in the last two years of that period. Equipment firms grew from four to twelve in 1984–97, and diagnostics firms from two to eight. Table 6.6 shows the distribution of biotechnology firms and support services.

Biotechnology is long established as basic science in Cambridge since this is where Crick and Watson discovered DNA in 1953. Later, Milstein and Kohler discovered monoclonal antibodies (1975), but the science was not patented in the UK, instead US biotechnology firms were founded for development of the technology to exploit these discoveries. The UK government was determined to encourage a cultural change among UK scientists to commercialize such cutting-edge discoveries on-shore, rather than allowing others to benefit from Britain's tradition of dispassionate pursuit of science as a curiosity-driven rather than commercially-driven activity. This was particularly strongly acted upon by the Medical Research Council, in whose government-funded laboratory the Milstein and Kohler discovery took place. One of the earliest examples of spin-off came from corporate rather than academic exploitation. Chiroscience started off in an incubator near Cambridge and is now Britain's strongest biotechnology firm, having merged with Celltech and acquired pharmaceuticals firm Medeva. The owner, Chris Evans has become Europe's leading biotechnology venture capitalist through Merlin Ventures, linked with Rothschild in its new DM500 million fund in Germany. Many new firms have spun off from academia, hospitals and laboratories. The Medical Research Council Molecular Biology laboratory has twelve spin-offs, among which Cambridge Antibody Technologies is one of Britain's fastest-growing biotechnology firms. There were, by 2000, around fifty core biotechnology firms in Cambridge and a further 150 in support businesses from supplying reagents to patent law and venture capital.

Hence, Cambridgeshire has a rather diverse biotechnology processing and development as well as services support structure, even though the industry is relatively young and small. Some of the service infrastructure, notably specialist lawyers and venture capitalists, and the scientific instrument sector, benefits from the earlier development of Information Technology businesses, many also spinning out from university research in Cambridge. Indeed, Cambridge's ICT industry is quantitatively greater than biotechnology, as we have seen, and also functions as a Working Cluster as will be shown below. The infrastructure support for biotechnology in and around Cambridge is substantial, much of it deriving from university and hospital research facilities. The Laboratory of Molecular Biology at Addenbrookes Hospital, funded by the Medical Research Council; Cambridge University's Institute of Biotechnology, Department of Genetics and Centre for Protein Engineering; the Babraham Institute and Sanger Institute with their emphasis on functional genomics research and the Babraham and St John's incubators for biotechnology start-ups and commercialization, are all globally recognized facilities, particularly in biopharmaceuticals. However, in the wider Cambridge region are also located important research institutes in agricultural and food biotechnology, such as the Institute for Food Research, the John Innes Centre, the Institute of Arable Crop Research and the National Institute of Arable Botany. Thus in research and commercialization terms, Cambridge is well placed in biopharmaceuticals; and with respect to basic and applied research, but perhaps less so commercialization, agro-food biotechnology also.

Within a 25-mile radius of Cambridgeshire are found many of the 'big pharma' or specialist biopharmaceutical firms with which commercialization development by smaller start-ups and R&D by research institutes must be co-financed. Firms present include Glaxo Wellcome, SmithKline Beecham, Merck, Rhône-Poulenc Rorer, Hoechst Pharmaceuticals (in 2000, the first two became Glaxo SmithKline and the last two Aventis through mergers). In the specialist biopharmaceutical sector, firms such as Amgen, Napp, Genzyme, Bioglan and Chiron are in the vicinity. Thus on one of the key criteria for successful cluster development, access within reasonable proximity to large customer and funding partner firms, Cambridge has this spillover advantage.

Finally, with respect to agro-food biotechnology, Rhône-Poulenc (Aventis), Agrevo, Dupont, Unilever and Ciba (Novartis) are situated in reasonably close proximity to Cambridge. Hence the prospects for linkage, though more occluded by public concerns about genetically modified organisms than in the case of health-related biotechnology, are nevertheless propitious in locational terms.

Cambridge is relatively well blessed with science and technology parks, though the demand for further space is significant. At least eight of the biopharmaceuticals firms in Table 6.6 are located on Cambridge Science Park itself. St John's Innovation Centre, Babraham Bioincubator, Granta Park, the Bioscience Innovation Centre and Hinxton Science Park were all expanded, completed, under construction, or under planning review in 2000. Most of the newer developments were placed within short commuting distance of Cambridge itself, on or near main road axes like the M11, A11, A10 and A14. This is evidence of the importance of *access* for research-applications firms to centres of basic research, reinforcing also the point that not everything concerning biotechnology must occur 'on the head of a pin' in Cambridge city itself.

The final, important, feature of the biotechnology landscape in Cambridge and the surrounding Eastern Region is the presence of both informal and formal networking between firms and research or service organizations and among firms themselves. Cambridge Network Ltd was set up in March 1998 to formalize links between business and the research community, connecting both from local to global networks in a systematic way. It is mostly ICT-focused, though some of this spills over into biotechnology, given its demand for instrumentation and venture fund investment. Of more direct relevance to the biotechnology community are the activities of ERBI. This biotechnology association is the main regional network with formal responsibilities for a newsletter, organizing network meetings, running an international conference, website, sourcebook and database on the bioscience industry, providing aftercare services for bio-businesses, making intra- and inter-national links (e.g. Oxford, Cambridge MA, San Diego), organizing common purchasing, business planning seminars, and government and grant-related interactions for firms.

Thus it is relatively easy to see that the Cambridge biotechnology sector operates as a cluster. Indeed, it could be said to be a paradigm case of the clustering phenomenon that, though presently small, has major growth potential. This is because it is Europe's leading biotechnology cluster in a business where the expected global turnover was $70 billion in 2000. It has the greatest non-US concentration of biotechnology drug development firms and these are the ones that will lead the industry in future. Germany, the nearest competitor, as will be shown, has concentrations at Munich, Heidelberg and Cologne but these are so-called 'platform technology' firms in important but internationally highly competitive areas like screening and diagnostics. Because of the sunk costs associated with co-location by venture capitalists, specialist patenting, legal, accountancy and insurance services, the immobility of the university as a key knowledge-driving resource, and the presence of a critical mass of biotechnology firms and entrepreneurs, Cambridgeshire is likely to remain the focus it has become.

These two cases are interesting and instructive for the light they shed on the leading role of small university or research laboratory start-up firms in leading a new industry and attracting large funding sources like the science foundations (in 2000, approximately $1 billion in Cambridge, MA and £1 billion for biosciences research in the UK) and big pharma to propel the industry forward financially. Because of the highly refined knowledge-base being exploited and the spillovers that take place habitually from close interaction between science and business, the cluster is the organizational form best suited to smaller firm development and growth. Cambridge, Massachusetts, is at least ten years older than Cambridge, UK, as a biotechnology cluster, but both show similar growth dynamics. Thus start-ups gather in incubators near the science base. As Zucker *et al.* (1998) show, there is a 'star' effect in that the presence of Nobel Prize winners is an important attractor of business as well as research interest. Thus Cambridge's MRC Molecular Biology Laboratory which has had ten such laureates, has become a key source of spin-out businesses, one, the aforementioned Cambridge Antibody Technologies, saw a thirty-fold share-price increase, made two significant alliances and raised £100 million in finance in March 2000 (Pilling, 2000). Meanwhile, Millennium Pharmaceuticals of Cambridge, MA, founded in 1993 and yet to generate a profit, had a market capitalization in the same month of $6.4 billion, acquired fellow Cambridge firm LeukoSite for $635 million and signed a deal with Bayer to develop 225 drug targets ranging from cancer to HIV. The conclusion for these firms must be that being in the cluster is not doing any harm to business.

New economy clusters: market-led

The examples examined here include the ICT industry in Cambridge, UK. Results from research based on linkage analysis interviews are detailed. Then, attention switches to the media and new media industries in Soho, London, SoHo, New York and on a smaller scale but also researched in detail, Cardiff, Wales. The focus in each case is upon industries where there is not a significant public research subsidy of the kind found in industries dependent on big science or military research budgets. These clusters work largely through networks between a variety of business and other appropriate actors who are familiar with each other's expertise, trustworthiness, reliability and willingness both to share relevant assets (e.g. information or lending a machine or employee if needed) and engage in normal business relationships based on market exchange. As we saw in Chapter 4, networks can be formal or informal, soft or hard (i.e. contractual, with an agreed project and business plan). In high technology industry, such linkages are more likely to involve research organizations such as universities both for knowledge, but also indirectly through spin-out firms. But once established, such links may be less pronounced than market transactions plus reputational trading and installed social capital. Aspects of this appear in Figure 6.2 which analyses key parts of the Cambridge ICT cluster, showing regular, co-operative as well as competitive relationships within the cluster.

Figure 6.2 shows a range of interactions for a small sample of randomly selected firms in Cambridge. At the higher added value end, interacting with the university and the Cambridge Network Ltd. are the top of the market software and systems design services firms like Cambridge Advanced Electronic (CAE) and Symbionics, itself linked to the Symbian hand-held devices consortium involving 3Com, Nokia, Psion and others. The systems houses have their own spin-out histories with larger established consulting firms like PA and Cambridge Consultants. It happens that CAE was in a design partnership with Tavisham and Minitura, called Synergy, but in 1999 a decision was taken to establish a virtual firm with individuals freelancing software code from the Eastern region, Turkey and Russia, still supplying large concerns like Shell,

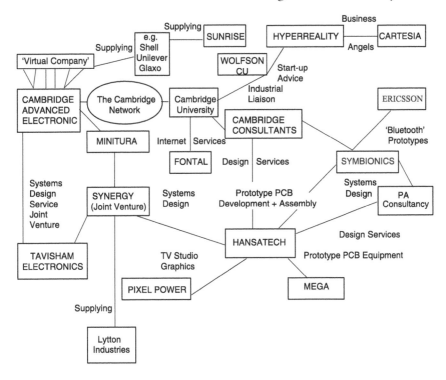

Figure 6.2 Aspects of the Cambridge IT cluster

Unilever and Glaxo but from an overhead-free environment, in fact the network leader's home in the Cambridge suburbs.

In a different corner is Internet games start-up Hypereality. It has a more tenuous link to the cluster because it belongs both to a new sub-cluster of like-minded firms and they have a counter-cultural perception of themselves in relation to the more sober image of ICT businesses. Nevertheless, Hypereality benefited from links with the technology liaison agency of the university and the Chamber of Commerce in accessing business angel financing from the Cartesia network. Previously a website design firm Hypereality had subcontracted to Fontal, an educational database software firm based in St John's Innovation Centre. Finally, in the lower right-hand corner is a group of firms collaborating on a 'Bluetooth' project which enables mobile telephones to communicate with other appliances like refrigerators (to see if they need re-stocking or to provide a recipe based on its contents). Ericsson were the prime contractor to Symbionics who contracted to Hansatech for the mobile telephony mother board assembly, itself a purchaser of special ultraviolet light-resistant printed circuit board substrates from Mega, also a supplier to university research laboratories. Hansatech had, in the same workshop, an on-board Formula 1 motor sport computer assembly project and had recently completed the graphics for a TV studio design for proximately located Pixel Power.

These are reputational market transactions based on trust in networks of firms and people known to be capable of performing tasks effectively and to extremely high standards on a global and local basis. They are the externalities on which firms rely to be able swiftly to meet tendering requirements and then fulfil exacting project needs.

Often, firms have owners or employees who are familiar with partner firms from having worked for them or known their key personnel as students. Hence, although only a segment or partial cross-section of the cluster in ICT in Cambridge, Figure 6.2 illustrates the kinds of links that operate inside the 'black box' and which cannot be accessed in any other way than by close and textured research. What are revealed by this approach are the evolving relationships among circles of partnerships and business or social acquaintances in the cluster, making use of social capital built up over many years and gaining enhanced, even global competitiveness accordingly.

Moving to new media, Nachum and Keeble (2000) show that London has at least 70 per cent of UK employment in media and 90 per cent in music. Of this, Soho captures about 70 per cent and, as they say: 'The entire chain of production – film production and post production, film distribution and sales agents, design, photography, music, advertising – is available in an area of about one square mile' (ibid., p. 11). Soho is also home to foreign media firms as well as those from the UK. Hence multimedia in London has a bias towards traditional media and advertising, growing from the audio, video and print media in the neighbourhood. The main kinds of new media market served involve corporate presentation, entertainment and leisure, particularly Internet games, and advertising and marketing, notably development of Internet websites. In a study by Russ (1998) he shows that London also has sub-clusters in Covent Garden, Clerkenwell and Shoreditch. The first-named is well-known as one of Europe's leading urban cultural zones, the ambience of which was reinvigorated when the capital's fruit and vegetable market was adaptively re-used for shopping and leisure activities to complement its traditional function as one of the world's leading places for opera. The other two are in London's East End, former industrial and working-class zones with Clerkenwell once London's clock-making quarter (Castells and Hall, 1994). These locations are selected because they are in cheap rent areas, yet the EC1 postcode is seen as valuable due to its proximity to central London's huge financial district. Much of the market for these firms is governed by the variety of clients locally available, especially in business services. Although far smaller than the Soho clusters, firms in these mini-clusters congregate for the same reasons, notably that it is useful to be co-located as an insurance against breakdowns of equipment or for sub-contracting. Trust in each other is highly-valued and accessed as a normal business 'spillover'.

In Heydebrand's (1999) and Pavlik's (1999) studies of multimedia in New York, the entertainment and financial sectors are powerful drivers of the industry with both functional and geographic proximity key features, albeit, as in London, the precise streetblocks occupied by the corporate giants are somewhat different from those of the new media firms themselves. In south Manhattan, also known as 'Silicon Alley', especially around SoHo and the iconic 55 Broad Street, where the New York Information Technology Center was founded, new media business involves entertainment software, Internet services, CD-ROM title development, and website design. Surrounding them functionally and spatially are the advertising, marketing, entertainment, education, publishing and TV, film and video producers and publishers. Heydebrand (1999) sees a three-level structure to the industry with the corporate giants resting on the mass of new media businesses who themselves relate importantly to Silicon Alley's creative and innovative capability as represented particularly strongly by 55 Broad Street. Because the last-named is unique to New York, it is more interesting to explore its existence as a public–private intervention than to repeat the locational linkages and rationales of the 1,000 plus start-ups in and around SoHo. That it functions as a cluster with horizontal and vertical networks based on trust, exchange, co-bidding and competition on contracts, sub-contracting and know-how and equipment-sharing as in London, is made clear by Heydebrand and Pavlik (1999) in their accounts.

55 Broad Street is different in that it is the product of the new economy governance practice of 'associationism' (Cooke and Morgan, 1998). With its first meeting in mid-1997 the New York multimedia innovation network set a mission of developing the city as the global capital of multimedia. It united mayoral and gubernatorial representation, Columbia University's NSF-backed Engineering Research Centre in multimedia, the finance and services community, venture capital, leading multimedia start-ups, major multimedia companies, arts and culture groups and industry associations. They met in the New York Information Technology Center, 55 Broad Street, brainchild of William Rudin, developer, who invested $40 million in superwiring the building as an incubator for multimedia start-ups. Technologically, the building offered, in 1997, turnkey multimedia transmission with satellite access, desktop fibre optics, C5 copper, video conferencing and 100 mb per second Internet bandwidth. There were 72 tenants in 1997, including International Business Machines and many start-ups in design, reportage, security and research among others. The association, like other cluster governance networks (e.g. the Massachusetts Technology Collaborative or the Massachusetts Biotechnology Council), has policies to lobby for tax, equity and rent incentives, business support programmes and expanding incubator opportunities.

Finally, to underline the importance of the link to old media in the evolution of new media in a part of the UK with its own language and media industry, we turn to the cluster in Cardiff, capital of Wales, one of the UK's four constituent countries. Deregulation in the 1990s stimulated the rise of independent TV programme producers, who, subsequently, stimulated directly through spin-off, and indirectly as part of a growing market, the formation of multimedia activity. In Figure 6.3 the cluster structure of the small firm part of the industry is shown while the organogram of the wide variety of private and some public actors that act as the enterprise support infrastructure for the cluster system is

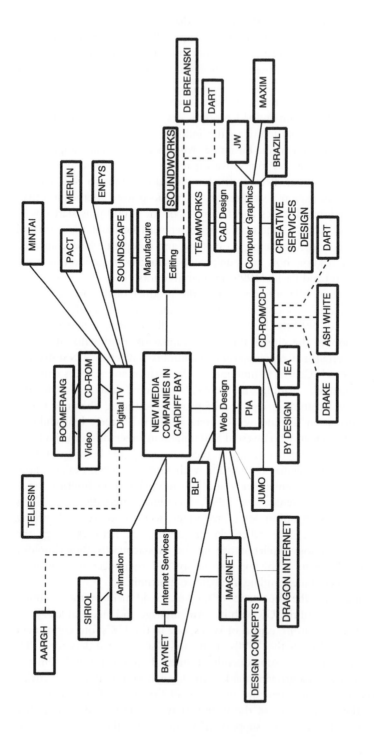

Figure 6.3 Selected new media companies in Cardiff and Cardiff Bay

described next. These include provision by TV broadcasters like BBC and S4C of scholarships and bursaries for different kinds of human capital, a public–private partnership of those two broadcasters and the commercial channel HTV with the Welsh Development Agency 'Screen Wales' representing the industry at Cannes, Venice, Los Angeles and other major festivals. Pump-priming or assisting in small firm bootstrapping through helping with accessing private, EU and Lottery finance for project development is also backed by this association to try to ensure that a fledgling film as well as new media industry, dependent mainly on a regional market but pushing to sell more widely, evolves. Cooke and Hughes (1999) provide an account of the industry derived from a survey of 300 firms in multimedia and media, animation, graphics, creative arts and services firms, some 16 per cent or 48 were core multimedia producers (on-line or off-line producers). Of the 145 firms responding to the questionnaire, 19 per cent or 28 were core producers. Of these firms, seventeen were specifically set up to produce on- and off-line products such as CDs composed of film and TV clips for entertainment; CD-interactive musical instrument tutoring; Geographical Information Systems; financial trading CD databases and media business CD data-bases along with vocational training CDs. A further eleven were providers of Internet Services or Web Page designers. Three of the first group and six of the second are currently located in Cardiff Bay. This is a waterside harbour area where relatively cheap and available space could be rented. It is somewhat comparable to the kind of former industrial, warehousing and office accommodation used by New York and Toronto new media firms showcased in the collection by Braczyk *et al.* (1999).

Turning to the Cardiff cluster, we see groups of firms that focus on specific market niches like editing and the manufacture of editing equipment, computer graphics, CD, Web Design, Internet Services Provision, animation and digital TV. There is overlap between old and new media activities among the last two groupings, but a supply-chain relationship between the others that meet the technical requirements of the TV, film or multimedia programme producers. As noted by Cooke and Hughes (1999), many firms cluster in proximity in the waterside area and gain spillover effects from so doing. Joint working on bids for project contracts is the most valuable of these effects and the possibility of more informal meetings by chance or by design is highly valued by creative specialists. However, not all firms in this mix of video, audio and text media like or need to be located in close proximity to the waterside cluster. Downsizing among dominant broadcast TV firms has meant the in-house demand for studio use and even physical space has been reduced. Sub-leasing of space to start-ups that find advantage in being located near to studio facilities means that this has helped to form two further, smaller sub-clusters in the city. Also a few of those firms surveyed did not seek to be located in the cluster but outside it although within reasonable access to the market, primarily the local one. The reason for not seeking proximity to other creative independents is a fear of losing a specific niche market should other firms learn about it. The possibility of this occurring anyway was perceived of as less of a threat than being in the milieu and, perhaps, speaking or being spoken about inadvertently. Such firms are few and apprehensive of competition, perceiving only negative externalities from being in close proximity to competitors in the cluster but advantage from being at not too great a distance from a client or clients. Finally, it is worth recording that in this study, when asked to rank the opportunity for networking with other like-minded firms in a closely proximate cluster setting, most firms ranked basic business imperatives like cheap rents and good security higher. Nevertheless some 20 per cent of firms ranked interaction of the kind clusters offer, higher than basic business requirements and these were overwhelmingly the creative part of the sample as distinct from the equipment or services supplier firms.

Conclusion

This chapter has been devoted to an exploration of a variety of 'new economy' industries such as telecommunications, biotechnology, ICT and new media that all engage in clustering activity. Starting with a review of ideas on the nature of the new economy by contrast with more established or 'old economy' industry, it was concluded that while there is no evidence that such industries have conquered either the business cycle or the need for stock market valuation to represent some measure of profitability as distinct from expectation or hope of profitability, nevertheless the new economy has some distinctive features. Among the more salient of these are a culture of tolerance, openness to new entries, recognition of the imperative to re-tool individually and company-wide in workforce skills and engage in lifelong learning, extreme competitiveness in production and consumption, the latter of a kind that challenges the consumer behaviour of the 'robber barons' of the nineteenth century, high interest and expectation of compensation through stock options and expectation of high job mobility and entrepreneurship tempered by confidentiality contracts regarding privileged knowledge. Most tellingly, each of these industries has a myriad small and medium-sized firms interspersed on occasions with large firms that might be the origin of spin-offs and the localized growth process, and may have moved into the cluster to capture spillover effects or may, even if located at some distance, be important to the cluster as a client firm.

This adds a dimension to the analysis of Norton (2000) to the effect that the key to the new economy is entrepreneurship formed by the meeting of scientists and engineers with venture capitalists and business angels. We have added the role of large public budgets in defence and science and large firms as clients of small specialists. As noted, large customer firms may locate or be already located in what becomes a cluster, but they may also be found at a distance, engaging in functional rather than geographic proximity. The case of Massachusetts-based Millennium Pharmaceuticals with its contract for 225 drug targets for German 'big pharma' firm Bayer is a case in point.

But it is more a norm than an exception in biotechnology. Indeed, as we shall see in Chapter 7, such has been the appetite of European big pharma for contracts with or part-acquisitions of American biotechnology and ICT firms that it is seen in some quarters as having contributed to the relative weakness and delayed development of European firms in these fields. The view of Bayer and others is that if such firms had existed in Europe, they would have made deals with them, but they were not to be found. This is an interesting debate within industry politics which may be instructive for the future, since one riposte to the large firm view is that they were often influential in advising concerned governments not to back new industries like biotechnology since pharmaceuticals was a chemically, not a biologically-based industry.

Notwithstanding these and related issues, such as the generally overweening influence of large firms in protecting their sometimes less than advanced perceptions of the future, or a related issue raised by Casper *et al.* (1999) to the effect that some countries, of which Germany is the exemplar, have a corporate institutional structure that militates against disruptive innovation, what do we learn about the nature of contemporary economic growth and development from the cases illustrated?

In Chapter 1 emphasis was placed upon the importance of recognizing the fundamental feature of industrial capitalism that is its disequilibrated nature. Marx, Schumpeter, Myrdal and Hirschman among others saw this clearly as a defining characteristic of market economies. For a long time these localization economies, giving rise nowadays to both agglomerations and, in more sophisticated ways, clusters have tended to be seen as mainly urbanization economies, something Krugman (1995) of the new neoclassicals tends to do. But while larger cities clearly contain clusters, not all clusters are found in large cities. In the knowledge-based part of the economy, centres of important knowledge-generation like Cambridge, UK, are within the broad orbit of a large city like London but increasingly self-sustaining for accessing the kind of marketized innovation support services necessary to finance or legally protect discoveries and applications. Even if part of an urban complex like Richardson or the Thames Valley, the cluster or agglomeration may give definition and meaning to an otherwise suburban or ex-urban sprawl. Then, new media businesses are seen to be thriving in the funkier parts of town rather than either suburbs or city centre. These illustrations capture the basic disequilibrium at the heart of economic development processes. Some locales have the necessary knowledge, social capital and contacts to create competitive advantage and those that do not have these assets are fundamentally hampered in their abilities to engage.

This issue, and its determinacy, are the subject of Chapter 7. Because, at one level, new economy industries and the clusters on which they seem generally to depend are so geographically uneven in their existence, national governments will always find it politically impossible to ignore the fact that they do not have a presence of note or at all in the industries in question. This concern was also present from the beginning of the Industrial Age and is inscribed in many forms from the statue in Ghent, Belgium, to the man who smuggled parts of the technologies driving Britain's textile industry forward in the nineteenth century to the failed attempts by different European countries to sustain independent computer industries in the twentieth.

Regional administrations too will want to do what they can within their spheres of competence to promote clusters and, as will be outlined, many have already implemented such policies. Thus the chapter returns us to the substance of Chapter 2 and the role of multi-level governance in the development of policies in support of innovation, learning and industry, in this specific instance in relation to development of new economy clusters, though more traditional forms are not ruled out. In essence, it is the cluster that is the primary object of interest, with the sector close behind. Or if it is the new economy sector that is uppermost, perhaps in national and supranational government thinking, then the cluster is the strong candidate as the preferred form, for reasons of competitiveness through enhanced productivity, innovation and new firm formation as discussed. However, there is evidence, as we have seen, to say that clusters are extremely difficult to will by public intervention and where the exercise begins from ground zero, they cannot be built. So we are exploring a dilemma of policy action. There is a desired outcome that evidence of the kind that has been reviewed demonstrates can be created, but there is no evidence that such an outcome can be achieved through design by policy. What will be done in the next chapter is to explore the extent to which the paradox just outlined is true and whether there is a role for distinctive levels of governance to act in specific ways to moderate the negative or stimulate the positive implications of the evidence that can be mobilized either way.

7 Can clusters be built?

Questions for policy

Introduction

In this chapter the focus is on the extent to which clusters can be built through acts of policy. To explore that important question it is necessary to take account of the conditions under which clusters currently exist. As we have seen, these vary considerably from old to new economy and in terms of scale and distribution. At the extreme, the paradigmatic case of a built cluster would be one started from ground zero by dint of policy intervention in the first instance. We have seen thus far many examples of clusters and none could be said to fit that category, though the role of public budgets can be shown to be crucial to the initiation and evolution of many, particularly where science or military funding is a key driver of research activity that results in commercialization. But even though public bodies may get involved by giving support of various kinds, the dominant cluster causation vector is one in which the firms that realize commercialization are mainly funded privately, often as risky ventures from which investors expect a generous return. That they do not always get it, but compensate the losing bets with the winners is the acid test of the thriving venture capital house or corporate venturer, as we have seen. But the earlier argument about the importance of proximity to other firms in the same or distinct but complementary sectors, as well as support agencies, suggests a degree of exclusivity regarding cluster potential. This is underscored sharply by the Schumpeterian thesis of disruptive innovation that causes swarming of imitators around the novel product or process in extraordinarily disequilibriated, even 'creatively destructive' ways.

These are circumstances of market success for cluster locations but market failure for the rest. Under such circumstances, as with the case for public subsidy of basic scientific research, public intervention is justified to seek to offset the failure. In the case of R&D appropriability, or the issue of whether investment results in a realizable market return, uncertainty is one of the key justifications for intervention. Where regional development is at issue, and clusters are a modern form of that, then justification for public intervention is harder in some settings than it once was. The UK, like other EU Member States is supposed not to subsidize existing firms because of prevailing competition rules operated by the European Commission. This was one element, though never taken the full distance, in the equation BMW was formulating when the decision was taken to sell its UK subsidiary Rover in 2000. However, for nascent industries, there is less opposition within the European Commission or the World Trade Organization, which has brokered agreements on industry subsidy between the trading nations and blocs it also regulates. It can thus be said that the theory underpinning the economic justification of public intervention for scientific research where the market fails has migrated to some extent into the sphere of industry policy (Arrow, 1973). This looks to be an important insight for policy-makers because it justifies support for clusters that are relatively easily presented as latent as well as scientific and technological in nature. Commercialization of science is the greatest impetus behind the new economy and its knowledge-based industries.

However, there is a snag deriving from disequilibrium in the distribution of sources in basic science, mainly universities, which are sufficiently close to the leading edge of research. Here the Information Age reaches its limits, for although websites represent the 'ubiquification' of information (Maskell *et al.*, 1998) high-value knowledge has a tacit dimension and this kind of knowledge capital tends to be geographically immobile even though the people who embody it may be anything but. Nevertheless, the critical mass of 'stars', their disciples, research centres, equipment and trusted contacts make for a remarkable immobility of leading scientific research centres (Jaffe *et al.*, 1993; Audretsch and Feldman, 1996; Zucker *et al.*, 1998). Where this does not occur, then observers are left to ponder their failed or stillborn clusters. Such are the circumstances under which building new economy clusters is very difficult, even in a country like Italy, which has strong clusters in traditional or old economy industry but none in the new economy. Thus Orsenigo (2000) points to the weakness in level and scope of basic research, university–industry relations, access to capital and unclear appropriation rights among the reasons for the failure of clustering to develop in Italian biotechnology, even in Lombardy its strongest region. Such an analysis is rare and enormously useful in confirming the importance of key drivers already discussed and suggesting policy lessons that can be learned both from 'success stories' which are mostly American and thus of limited direct adaptability, and other cases where policy adjustments or major interventions have improved a weak initial position.

In what follows, an effort will be made to summarize key reasons for a reasonably large number of both new, mainly small, and older economy clusters in the UK, a European country, but one that many observers equate rather readily with the USA due to their presumed Anglo-Saxon affinities in business culture (Casper *et al.*, 1999). Some comment is offered on the evolution of multi-level governance thinking in relation to cluster support policies in the UK, aspects of which are influenced by US, particularly Massachusetts, practice.

Then attention will briefly be devoted to some Nordic instances of successful cluster building around universities, one near the Arctic Circle in Finland, where the initial conditions of a declining tar industry were even less propitious than that of the declined textile base of another case further south in the same country. To a limited degree, European Union finding helped these developments but only lately and marginally compared to the major exercise in technology-focused public–private partnership and enterprise by national government operating in a weak regional governance context.

Finally, an in-depth account will be given of the major public policy to build clusters through the BioRegio programme, a federal initiative in Germany which relied fundamentally on capacity-building at regional and local levels in an initially unpropitious and still somewhat constrained institutional environment. Then there will be reflection back on the implications of these cases for a policy-learning approach towards dealing with disequilibrium that pays close attention to a multi-level governance perspective. One of the most obvious features linking these three cases of declining clusters is the absence of either strong regionalization or regionalism as a driver for associative governance practices in support of innovation and learning. Belatedly, multi-level initiatives and funding to re-new the cluster are being mobilized in the ceramics district centred on Stoke-on-Trent. The Nordic instances, as small, relatively unregionalized countries, rely on close interaction in innovation governance between knowledge-based localities and the national centre. But Germany shows best how past failure on a too centralized approach has been turned round by promoting multi-level governance of biotechnology innovation focused on its strongly *regionalized* structure of sixteen *Länder*.

Comments on decaying clusters

In Figure 7.1 the distribution of some of Britain's clusters is presented. These are picked out as having been the subject of studies that either were of these phenomena viewed through the lens of cluster theory or one that allows such an interpretation to be made. Some have already been discussed in Chapters 4 or 5, and readers are referred there for detail. In the main, the interest here will be in tracing the extent to which what exists is influenced by specific policy instruments and, of these, which seem to have been helpful and which not. The focus is initially on three older economy clusters in England where it could be said the phenomenon was first observed. An absence of regional economic development competences until 1999 when Regional Development Agencies with nominated chambers were established, reveals some of the problems of too much localized control, on the one hand, and too much of the wrong kind of central government intervention, on the other. A third example involving partnership among all three levels, not discounting the EU as a fourth has better prospects. Cluster performance in the UK has been mixed with some limited efflorescence of new economy clusters, the further entrenchment and growth, fuelled by major foreign direct and indirect investment in London's financial cluster, but the probable weakening of at least one long-established one in the case of Birmingham's automotive industry and its supply-chain based cluster. Indeed, the automotive cluster centred on Birmingham in the West Midlands

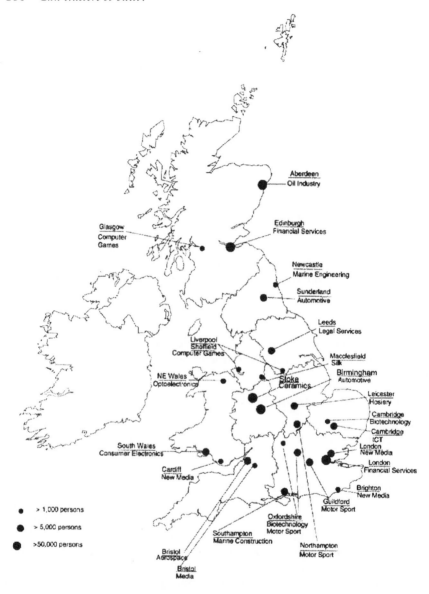

Figure 7.1   Selected industry clusters in the UK

will be the subject of the first exploratory dive into UK cluster practice because it demonstrates the severe limits placed upon external intervention of any kind, public or private, by what has become a mature cluster. It is one that was restructured in significant ways by national government intervention, first to nationalize and consolidate the main assembly company, then to privatize what had become the Rover car company. It is at approximately the latter point that some of the key issues concerning the culture of established clustering are explored in what follows.

An interesting insight into the distance between the power of rhetoric, and that of brute reality is provided by Marquardt and Sashkin (1995) in a chapter entitled 'Rover – one organization's journey to success as a learning organization'. Rover is of central, long-established importance to the vertically structured Birmingham automotive cluster. The importance of learning to the future of the firm was symbolized in quotes from chairman of the company in the early 1990s, Sir Graham Day: 'As a company we desperately need to learn' (ibid., p. 195) and 'Neither the corporate learning process nor the individual is optional. If the company seeks to survive and prosper, it must learn. If the organization and individual seek to make progress, learning is essential' (ibid., p. 193) and from Chief Executive, John Towers: 'We have achieved and will continue to achieve dramatic progress through people who are perpetual learners, people who are skilled in the application of that learning' (ibid., p. 193). At the time the firm was owned by British Aerospace, in a valuable 'learning' alliance with Japanese firm Honda and shortly to be sold, to Honda's surprise and disappointment to BMW in 1995 before the latter offered it in 2000 to UK venture capitalist firm Alchemy and finally sold it to the local Phoenix consortium in 2000. For a time it looked as though the cluster might be devastated if a buyer could not be found, and it remains likely that employment in the local and national supplier base will continue to decline. The lessons learned were bitter indeed.

When Rover set up the Rover Learning Business (RLB) initiative, four of the programme's five objectives focused inwards on learning in the firm and one outwards to the 'Extended Enterprise'. The internally focused ones involved enabling employees, or 'associates', to 'climb the learning ladder', providing learning process tools and techniques, conducting corporate benchmarking and achieving an image as 'world best-in-class learning company'. For suppliers, the aim was stated to be to offer 'learning support and collaboration to facilitate world-class activities'. This involved offering courses 'to help suppliers meet the auto industry's demanding standards of quality and efficiency' (Marquardt and Sashkin, 1995, pp. 195–204). Targets to be achieved included specified cost savings from learning, measurable shifts in employee attitudes, employee self-development and career motivation, managers to act as qualified coaches 'as opposed to cops' (ibid., p. 197), and large-scale external accreditation. All this was to be set within a 'lean management' framework, then being widely adopted following the publication of the book that explained the success of Toyota in global markets (Womack *et al.*, 1990). It would not be exceptional for automotive firms in many Western countries to be going through programmes like this, and some are reported as having distributed multiple copies of the book in question to their managers in the early 1990s (Cooke and Morgan, 1998). The question in Rover's case that is of particular interest here is what effect the RLB programme had, not so much internally, as externally, upon suppliers.

Some limited evidence on this and a specific, important supply chain comes from a UK research council and Lucas co-funded study (see, for example, Towill, 1997). Rover's first tier supply-chain relationships in its core region had been described as a 'stress' management relationship as early as 1984 in a report commissioned by the West Midlands Enterprise Board from a group (Bessant, *et al.*, 1984) containing one of the co-authors of the Womack *et al.* (1990) study. Elliott and Marshall (1989) wrote of the cluster having 'been trapped in a downward economic spiral as its former strengths in interlinked, diverse metal-based manufacturing have become structural weaknesses' due to a history of under-investment, technological backwardness and a collapse in skill training (ibid., p. 228). Investment was transferred overseas to escape labour conflict and indigenous demand for machine tools declined by 70 per cent. The crisis in the cluster nearly brought it to an end, but the privatization of what had become a nationalized industry and the recruitment of overseas management like the Canadian, Sir Graham Day to replace the South African Sir Michael Edwardes to run Rover was part of the recovery strategy with 'organizational learning' by now a key instrument. However, the Lucas study showed three types of supply-chain culture characterizing the Lucas relationship with its customers and suppliers. They are each means of withholding information known as 'over to you', 'badging' and 'we know best'. Each is profoundly at odds with concepts like networking and clustering. The first of these is typical of the relationship with Rover. The 'over to you' syndrome means a supplier wins a year's contract to supply a large number of components but the customer refuses to forecast a weekly breakdown, saying 'just deliver what we want when we want it' (Towill, 1997, p. 37). Qualitatively, a senior Lucas manager described this as the customer putting in an order on Thursday for delivery on Monday, changing the specifications maybe six times in-between. In other words, thirteen years later Rover retained its 'stress' management relationship with its most important supplier unchanged.

This approach naturally produced repercussions down the supply chain. Within Lucas, this enhanced an already cut-throat philosophy in dealing with suppliers, whether external or internal to the company. Thus the automotive purchasing department consisted of some fifteen managers in competition with each other, negotiating all day by telephone to drive supplier price down as far as humanly possible. A bonus scheme rewarded the winners. In one particular internal supply chain plant, fluctuating but generally rising demand for a key engine component was resulting in failure to meet orders. The syndrome here is close to 'we know best', where an intermediary places different cyclical time-pressures on a supplier than those emanating from the customer. This failure was triggered by the presence of an intermediary plant located elsewhere that received customer orders, production of which, however, it was neither responsible for nor competent to carry out, but to which products were then dispatched for making up into mixed consignments to be sent to the customer. The intermediary also held customer information, updated it but passed it to the component plant infrequently. Staff morale was, not surprisingly, low and staff turnover was accordingly high, leaving a shortage of trained operators and inadequate craft and supervisory support. Recruitment of untrained personnel, of whom as many as 25 per cent would leave or be rejected weekly due to inadequate skills, led to a high production of rejects and a scrap rate of 50 per cent, largely because of operator inability to operate machines correctly. Accordingly, productivity at the plant was low.

An independent Lucas supplier in a nearby location with higher productivity, a more stable workforce and a role as both customer of and supplier to large Japanese companies had adopted work action teams for problem-solving on the shopfloor, a customer focus which pervaded and gave purpose to the plant, a business unit structure allowing clear performance monitoring, a quality culture 'pushed by continuous improvement and the high standards required by Japanese customers' and 'an apparent trust atmosphere that has resulted in a good working relationship with the trade union' (Lewis *et al.*, 1998). This plant also benefited from close location of suppliers and frequent deliveries. These are intertwined in their positive effect and offer a clear contrast with the previous example where the opposite conditions prevailed. Thus frequent small batch deliveries could be made and close supplier location improved communication and assisted quick problem resolution. Finally, unlike its low productivity comparator the second plant had a low scheduling variability from its customers, aided in some cases by electronic data interchange, enabling accurate demand forecasting and materials planning.

The key points to keep in mind from this account of the ways in which comparable plants can experience wholly different business trajectories depending on the nature of final customers are as follows. First, while the key customer to the first supplier professed itself to be a 'learning organization' with a commitment to offering 'learning support' to the 'extended enterprise' including suppliers, it had not, roughly a decade later, begun practising what it preached, at least in relation to one of its most important and long-established suppliers. This had a knock-on effect of forcing stress down the supplier's own internal supply chain. Thus attempts to produce cost-savings by hierarchical forms of communication and consolidation of distant and distinctive consignments had produced the opposite in the forms of high scrap and low productivity rates. The firm exposed to true learning organizations had absorbed that culture, was making use of cluster effects from geographical co-location not for their own sake but because they enhanced the achievement of business goals, the main one being high quality customer service.

Second, the absence of trust, other than in the market mechanism, was pronounced in the first case where, perversely, resources were being wasted because it was assumed new employees would be able to perform complex tasks without the need for any training, a paradigm case of 'misplaced trust', it can be deduced. Trust was sufficiently high in the comparator plant for trade union involvement to be integral to its successful functioning, and crucially, an active training policy meant 75 per cent of the workforce was multi-skilled, giving greater efficiency and flexibility to staff deployment and more varied career opportunities to employees.

Third, location enters the equation of greater and lesser plant accomplishment as a necessary facilitator of heightened efficiency. Location close to customers and suppliers, difficult to achieve in the absolute, is the ideal when firms operate as learning organizations. When they do not, what are being discounted are locational spillovers such as those which make problem-solving interactions easier, or facilitate good, clearly understood communications as key means of optimizing customer satisfaction. This is because the cost-reduction culture has made it difficult to develop the quality culture practised by successful modern firms. Once that has happened, energy is focused on cost-saving as a survival strategy for the firm. When incompatible locational and organizational structures are the focus for attempted improvements, this can lock in the firm to making decisions that increase inefficiencies, like hierarchical communications with low transparency, rather than the efficiencies being sought. From an evolutionary economics perspective, the selection mechanism has limited scope because learning capabilities have been insufficiently developed or implemented.

In other cases of declining clusters in the UK, ignorance or dismissal of locational advantages that can be enhanced from proximity have been highly damaging. The silk manufacturing district of Macclesfield still retains some clothing production specialization but silk hardly features in the way it once did. Historically, silk was imported from China and woven for women's dresses, including shot-silk and satin, with important sub-contracting linkages to firms specializing in embroidery, gold wire embroidery being a speciality. Other markets included military materials such as flags, and for the Royal Navy and beyond, silk scarves. In wartime, maps were embroidered in silk to preserve them and even silk cigarette cards were produced. In a study of some of the poor decision-making by owners of silk weaving firms, Pyke (2000) notes a not unusual imperative driving their main business concerns. This was to keep wages low by keeping out alternative sources of employment. Control of local government industry policy through political influence ensured this did not happen until the 1960s, when the industry had ossified and ICI (subsequently Zeneca) and Swiss pharmaceuticals firm Ciba-Geigy (now Novartis) were allowed to locate in new industrial estate property in the town. Umbro the sportswear manufacturer, once a user of silk for boxer's shorts and dressing gowns, also moved in.

Second, it is noteworthy that manufacturers never sought to move up-market into more value-added production, blaming control of markets in the UK on London merchants who, instead of pushing the Macclesfield industry towards learning new processes or skills, simply consigned UK silk production to commodity status and purchased quality fashion silkwear from Italy's silk district of Como. Accordingly Macclesfield producers sought to maintain their position by moving in a different direction, towards man-made fibres, especially the lower price silk-substitute rayon. Third, Macclesfield has remained a reason-

ably prosperous industrial town in which nowadays a few large employers dominate the labour market. Umbro has grown considerably and supplies to a wide
range of sportswear customers, not least in the burgeoning market for up-scale
football kit. However, there is no longer a strongly localized cluster operating in
ways it once did, with a substantial amount of local sub-contracting, although
some smaller design firms are still found connected to the industry. Given the
protective assumptions of employers towards 'their' labour surrounding early
'Industrial Age' industries like silk manufacture in the UK, it is difficult to see
how policies could have been designed to help modernize and even upgrade the
industry. On the contrary, when, finally, the newcomer industries were let in,
they contributed to the future relative prosperity of the local labour force while
hastening the demise of the cluster whose firms found it still harder to compete
while bearing higher wage costs. The route taken was for ownership to concentrate, production to be automated and scale economies to be sought in new,
artificial fibre production.

Something not wholly different overtook the considerably more important
Potteries district centred on Stoke-on-Trent. In recent years authors examining
the fate of this once leading cluster have commented on the damage done to
the UK's ceramics industry, particularly fine bone china production by the
belated attempt to gain global production and marketing scale through mergers
and acquisitions (Imrie, 1989; Padley and Pugh, 2000). These have effectively
disconnected the industry from its locally embedded culture of family ownership and localized linkage through sub-contracting for specialized inputs and
outputs as described in Machin and Smyth (1969). The industry moved from a
classic position theorized by Stigler (1951) where:

> an earthenware manufacturer would build his own kilns; make his own kiln
> furniture; calcine and grind his own flint; make his own colours; frit and
> grind his own glazes, and carry on a whole range of operations that are
> now performed on separate premises by supplying industries.
>
> (Machin and Smyth, 1969, p. 5)

to one in which:

> manufacturers tended to become more specialised over time. Specialised in
> the sense that outside suppliers satisfy many requirements: prepared raw
> materials, colours, lithographs, machines, kilns etc. This reduces the size of
> the units which restrict themselves to making, firing, glazing and deco-
> rating.
>
> (ibid., p. 56)

This was the position going into the 1960s when the industry moved away
from craft organization, the linkages of which are shown in Fig. 7.2, to more
scientific management, shareholder-driven automation and more defensive as
well as offensive competition in new markets.

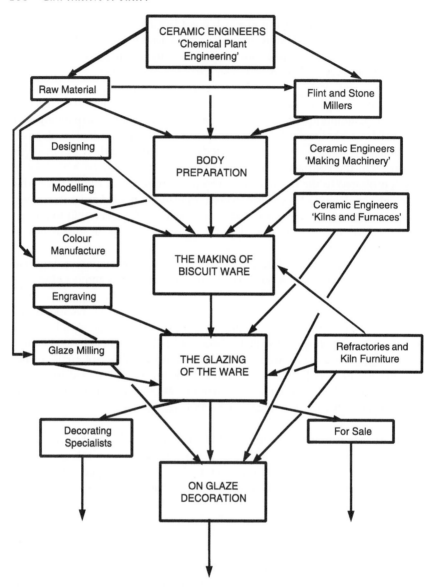

Figure 7.2   Selected ceramic cluster linkages in Stoke-on-Trent, 1930
Source: Machin and Smyth (1969)

In Padley and Pugh's (2000) study it is shown that at £2 billion the industry is nationally if not internationally significant but that output declined by 50 per cent between the peak post-war year of 1976 and 1999. The industry is cyclical but poorly planned at the firm level to cope with what can easily be seen to be a ten-year business cycle. Thus firms invest near the top, then cut back swiftly at the downturn, finding themselves short of capacity at the next upturn. Domestic demand increases are thus satisfied by imports. Both the firms responsible for Stoke's duopoly, Wedgwood and Royal Doulton found them- selves vainly seeking former employees made redundant a year earlier with a sharp increase in demand from the US for hand-crafted products in 2000. Sub- contracting to rivals at home and abroad, the course taken, comes at a price.

The duopoly accounts for one-third of industry output and more than half that for tableware, having been formed from amalgamating numerous indepen- dent firms in the 1960s and 1970s. Poor returns on investment since then led to Wedgwood to seek a 'white knight' takeover by Waterford Crystal in 1986 and Royal Doulton, previously a subsidiary of the media group Pearson, being floated as a public company on the London Stock Market in 1993. Waterford Wedgwood has a 15 per cent share in Royal Doulton and is on record as willing to make a full bid to keep outside competition at bay. Hence the imperative to further concentrate ownership retains a strong hold in the industry even though cyclicality and short-termism in the City have had devastating effects on both firms. This has been less the case for design-intensive firms like Portmeirion and Churchill, who are not under the same rationalization and downsizing pressures as publicly-quoted firms. One long-established business, Tams Group, went into receivership within twelve years of going public, having survived for over 100 years in family ownership. One difference in market strategy, which has helped the industry survive, by comparison with Macclesfield silk, is that fine bone china is the luxury speciality of the duopoly. Moreover, growth markets are in household ware and porcelain. The latter is hardly produced in the UK and dominated by the likes of Limoges, Dresden and Meissen, but the UK holds second position globally in terms of exports by value of ceramic house- hold ware, mainly because its output is such a high-cost product. Imports, however, have risen to 40 per cent, much of the increase from Thailand, China, Portugal, Germany and Italy, hence not only from low labour-cost countries.

Industry consultants advocate a broadening of export markets, establishment of a collective market research service the better to cope with the cyclical nature of the industry and reproduce local conditions whereby:

> a continuous stream of new entrants replaces companies that fail or are merged with others. This is how the ceramics industry was formed in the 19th and 20th Centuries with many famous names surviving. These include Spode, Wedgwood, Doulton, Wild, Colclough, Midwinter, Shelley.
>
> (Padley and Pugh, 2000, p. 23)

Few entrepreneurial firms that are typical of cluster settings were formed in the second half of last century and the two that came to prominence, Portmeirion and Denby, both started elsewhere. But mergers have produced poorer industry indicators, except value of exports, and significant rationalization of employ- ment. So at least some experts, backed by the British Ceramics Confederation conceive of a healthier future for this former cluster lying in localized policies to improve medium-term financing, new firm formation, flexibility in response to market shifts, greater market intelligence and forecasting, innovation and creativity, along with distribution support or even cooperative marketing of small firm production. If such a set of policies were to be implemented, it would mark a recognition of the importance to the future of the UK ceramics industry of building up new cluster connections in a locale that once had them and their associated economies of scope in an earlier form, but lost them in the mid-twentieth-century quest for economies of scale. As these experts note: 'In the same way that in the past foreign competitors learned from – indeed, in some cases copied – UK products, UK manufacturers now need to learn from their competitors' (ibid., p. 23).

Learning from competitors whether clusters can be built

It is clearly difficult to recreate the virtuous cycles of cluster involvement once lock-in to outmoded mentalities has occurred, and cognitive lock-ins which eschew a growth in favour of a survival perspective are harder to implement than pay lip-service to, as we have seen. But all need not be doom and gloom even here, as the case of the learning organization profiting from taking a customer-focused view, influenced by learning from Japanese customer require-ments shows. Instances such as these may have a valuable demonstration effect, something that occurred in the 1990s among German automotive suppliers who were encouraged to learn from change-managed companies by the main trade union, IG Metall as well as management and various government ministries (Cooke, 1996; Cooke and Morgan, 1998). To a limited extent some-thing comparable is emerging at a collective level in the UK ceramics cluster in Stoke-on-Trent, a Task Force has been established with UK, regional and local government partnership support and a Ceramics Industry Forum funded to identify best practice and disseminate it. Macclesfield's silk industry heritage has, as in many other areas with a pioneering industrial history, been turned into a cultural industry based on a number of museums, exhibitions and working replica mills.

One of the countries that took the task of building *new economy* clusters in older industrial centres in decline most seriously is Finland. From the early 1980s, a national strategy to build knowledge-based clusters in such fields as telecommunications, automation, medical technology, biotechnology and some fourteen others was centred on the unremarkable idea of establishing science parks at ten universities around the country. However, unlike the implementa-tion of this basic idea in many countries, seeking to replicate the model of Stanford Science Park and its subsequent influence on the growth of Silicon Valley, the Finnish version seems to have worked. The key reason for this is a policy of linkage between university research, R&D laboratories of large compa-nies such as Nokia or ABB, some of their suppliers, and start-up firms spinning out largely from university research. Typically, these are co-located on or near the technology park. Finance for start-up firms comes from project-based contracts they earn from these and other customers while seed-corn funding comes from private borrowings and some co-funding from national technology support agencies like the Finnish Technology Development Centre (TEKES). By the late 1990s the parks, all belonging to the Finnish Science Park Association (FISPA) contained around 700 spin-out companies and units employing some 7,000 people. TEKES and FISPA together host EU co-funded facilities such as the network of Finnish Innovation Relay Centres which disseminate results of EU-funded research for commercial exploitation, and assist in international partner-finding to enable EU Science and Technology (Framework Funds) and other research and commercialization funding bids to be filed.

An outstanding case of the success of this approach in what looks geographically to be an unpropitious setting is that of Oulu, a city of 100,000 people located at the northern end of the Gulf of Bothnia on the Baltic Sea, 230 kilometres from the Arctic Circle. Oulu's industrial history rested on the extraction of tar from pinewood and other forest product activities. The tar industry disappeared in the twentieth century and replacement industries like chemicals and wood pulp are highly capital-intensive. So to help sustain the community in such a remote setting the government established a university at Oulu in the early 1960s, setting up the key engineering faculty in 1965. This was followed by the electronics and automation branches of the State Technical Research Centre (VTT) in 1974, the Oulu technology park in 1982 and the medical science park in 1990. In parallel, Nokia first established a cable manufacturing plant in Oulu in the 1960s, then established its first advanced technology facility there in the 1970s. By 1996 there were six divisions of Nokia employing 4,700 on and off the technology park: cable employed 400, network systems 230, mobile phones 390, access systems 1,180, LK-Products (communications compatibility filters), 1,220, and cellular systems 1,280. Among Nokia's key suppliers attracted to the area are printed circuit board firm Aspo, metal cabinet-maker Scanfil, and electronics systems developer Elektrobit.

Excluding Nokia, the technology park had in 1996 1,200 jobs in some 120 firms (Jussila and Segerståhl, 1997). Most of these are engaged in research related to telecommunications ranging from software and sensors to optoelectronics and lasers generating an annual sales value of over $1 billion. The combination of three distinctive kinds of actor, university and public laboratories (research and training), multinational with suppliers (research and production), and start-up businesses (research and exploitation) supported financially by private venture capitalists such as TeknoVenture and public development investors and technology experts TEKES has been fundamental to the success of the cluster. Since 1990 the Oulu model has been applied to exploitation and commercialization of medical research with the establishment of the Oulu Medipolis, a technology park with 300 employed in some fifty firms 'operating close to researchers in order to create networks and interactive cooperation between industry, SMEs and research institutes globally' as the official brochure describes the mission (Lampulo, 1996, p. 16). Internationally high ranking collagen research at the University Biocentre attracted Californian rheumatoid arthritis drug company FibroGen to locate on-site in 1996, sharing core Medipolis laboratory facilities with other firms on a round-the-clock basis. Other, indigenous firms are active in biotechnology-derived treatments, diagnostics and biomedical devices.

Oulu was the first technology park development to be designed from scratch in the Nordic countries. It is clear from the experience of knowledge-based cluster development there, elsewhere in Finland, even in difficult settings like the former textile city of Tampere, and in Sweden as at Linköping, Gothenburg and Lund, that the combination described above works successfully in the creation of knowledge-based clusters. However, it is important to keep a number of caveats in mind regarding the exportability of a distinctively Nordic model of cluster development to institutionally very different policy settings. The first of these is that all the countries in question, but especially Finland, have long histories of corporatist governance in which public investment is a key developmental trigger. This works extremely well in most respects as the high standards of living, welfare and buoyant economic growth of recent decades testify. However, such countries can appear to have 'managed markets' in some respects, whereby strong public procurement and domestic loyalty provide indigenous large business with the means to generate the surpluses that allow aggressive competitiveness on international markets. Consensus for choosing a broad strategy, planning it in some detail and each partner playing the correct part are also institutionally key elements in this success and not easily replicable in different institutional settings.

A second point is that it is not entirely clear how competitive locally, let alone globally, start-up firms in the Finnish and other Nordic countries would be if they operated without the kind of protection and preferred supplier status they enjoy in relation to key customers. Clearly some firms can be fast growth 'gazelles' as Feldman and Klofsten (2000) show but other work by Lindholm-Dahlstrand (1997) shows that technology start-ups, even in the relatively benign context of Sweden, are generally slower and remain smaller if they originate from university research as compared to being corporate spin-off firms. In brief, many of the firms in question could be described as not being fully market-facing where they operate under 'infant industry' protection from customers like Nokia in Finland or Ericsson and Saab in Sweden. Finally, it is not clear how much such firms are and remain for some time dependent on public subsidy for services like training or accessing investment finance, something which adds to a sense of scepticism about their likely long-term sustainability. Thus Nordic universities subsidize entrepreneurship training for scientists and engineers setting up businesses, thereby contributing significantly to small firm survival rates. But, objectively speaking, all these policies make sense, especially as compared to the approach in the UK in the past which could best be described as 'sink or swim' entrepreneurship, with the regional imbalances implied by the concentration of venture funding in the London region present to make it even more unlikely that an Oulu event could occur.

It is instructive to see how the UK Labour government after 1997 recognized market failure in academic entrepreneurship and intervened with schemes to encourage it such as University Challenge and Science Enterprise Challenge. It is also noteworthy that in the home of free enterprise, the US government makes major public seed-corn funding available for research commercialization through its Small Business Innovation Research (SBIR) scheme and a panoply of state-level incentives. Hence it is not a case of no role for the public sector in cluster building, but under conditions where market failure frequently occurs because of investor risk-aversion, what is the best role?

Germany's BioRegio cluster-building strategy

German biotechnology excellence in research had never been matched by extensive commercialization activities and its multinational pharmaceuticals firms have their intellectual origins in synthetic chemistry. Despite a number of public initiatives aimed at promoting exploitation through the encouragement of business start-ups from universities and research institutes, it was not until the launch of the BioRegio contest in 1995 that the commercialization climate became favourable. It is widely understood that pharmaceuticals firms contributed significantly to Germany's weak commercial biotechnology sector by advising government officials against bioscience in favour of chemistry for reasons of ignorance and self-interest. The change was brought about by German Ministry of Education, Science, Research and Technology (BMBF) targeting funding under its Biotechnology 2000 programme to three regions which proved strongest in meeting required commercialization objectives. These regions are located in the Rhineland, centred on Cologne, Rhine-Neckar-Dreieck, with Heidelberg the key scientific base, and Munich. In all, seventeen regions competed for funding under the BioRegio contest, and as well as the three winning regions, Jena in former East Germany was given a special prize. The funding available for BioRegio is DM150 million over five years, allocated from the overarching Biotechnology 2000 programme.

Despite its relatively small size, whereby each winning region has some £3 million per year to support commercialization of biotechnology research, BioRegio has had an important symbolic and practical effect on the innovation system associated with biotechnology in Germany. More than any other federal initiative it has produced rapid, positive results and galvanized entrepreneurship in respect of new firm formation, also giving a significant boost to Germany's lagging venture capital industry. Hence, the policy warrants detailed examination not least because it departs from many previous initiatives by taking a regional approach to the promotion of industry and seeking to support and develop leading biotechnology locations as clusters, an approach which is based on the debatable assumption that clusters can be built by design. To the extent that positive results have emerged from this policy, the most important condition from the outset is that cluster-building has been promoted from a strong science and financial base rather than from a ground zero position. As with the

Nordic cases discussed earlier, in knowledge-based industry this is a necessary if not sufficient condition.

BioRegio represents an approach that goes beyond the normal role of government research funding by trying actively to build a critical mass of commercial biotechnology business enterprise through public intervention. It is notable how dependent the whole industry is upon varieties of public funding from basic and applied research to venture capital and in the long term it is not sustainable, but venture capital is expected to grow in abundance to take the place of subsidies removed in the process. By the time biotechnology start-ups need second round funding, between three and five years after receiving initial funding, public funds will not be capable of meeting demand. Thus it was expected that there would be a continuation and significant growth of the late 1990s trend for US and some UK venture capital to move into Germany substantially to augment the relatively small indigenous venture capital sector. This expectation was not unreasonable and growth has occurred generally and particularly strongly in Munich and the Cologne–Düsseldorf region. Nevertheless, it is also expected that second round financing will be accompanied by a phase of firm consolidation throughout Germany involving mergers and acquisitions, but also firm insolvencies. Caution led BioRegio managements to fund only quality projects with high market potential initially. Accordingly, failure rates were low, but even so, most new firms started in globally competitive platform technologies such as diagnostic kits where attrition rates are high, hence the expectation of consolidation.

Assessments of the impact of BioRegio reported by Dohse (2000) showed firms in the industry sharing a view that the policy was perceived to be successful in terms of inputs, both directly from the public purse and indirectly through stimulating a growth in venture capital funding. The number of biotechnology firms in Germany grew from 170 to 222 during the first two years of BioRegio's implementation, although Giesecke (1999) queries definitions, estimating the actual number as probably less than two hundred. By 1998 federal expenditure on R&D in biotechnology and molecular medicine had reached DM1.2 billion (£400 million) and *Länder* funding added a further DM0.8 billion (approx £300 million), making DM2.0 billion (£700 million) overall. Venture capitalists grew in number from two in the 1980s, namely Atlas Venture and Techno Venture Management, to more than seventy engaged in biotechnology investment by 2000. This shows how relatively effective were the assumptions of policy-makers that by the time demand for venture capital outpaced the capacity of the public sector to provide it, the private sector would have grown to meet likely demand. Of course, the nature of policy-making and implementation in this field in Germany meant that such expectations were not guesswork but effects of the consensual and partnership-minded process underlying implementation of the policy.

Firms from within and outside the winning BioRegios were surveyed by Dohse (1999) to discover their assessment of the policy. In all, 100 biotechnology firms from all seventeen regions competing for BioRegio status and from outside were asked about the innovativeness of German biotechnology, barriers to future innovativeness and the role of the BioRegio contest in influencing the competitiveness of the German biotechnology industry. The key aim was to discover whether the regional clustering assumptions of the policy were strategically efficient or not from the viewpoint of firms in the industry. Some 33 per cent of firms responded, thirteen participated actively in the contest, twelve received funding from the programme and eight neither received funding from, nor participated in the programme. Table 7.1 summarizes general findings on firm perceptions of the innovative condition and barriers in German biotechnology. The strongest support for the policy's key assumptions concerned the perception of a technology gap in German biotechnology. This was seen to be problematic both for commercialization and applied research. A majority of firms stated that it was at least partly true that BioRegio had helped reduce the technology gap. A further key assumption of the policy was that there was a marked lack of interaction between key players in the sector. This view was widely shared among firms responding, indicating the validity of the assumption and recognition by firms of their own experience of market failure. Firms were asked, among other things, for their general assessment of the BioRegio contest and the answers of all categories of firm are provided in Table 7.2. This reveals how much the initiative was perceived as something of a saviour of Germany's potential as a biotechnology competitor.

Some conflicting views offered by firms were: that future support should be on a national rather than a regional basis, and that innovative firms should be able to link to BioRegios even if located outside them. There was a view that start-up firms should be strengthened in relation to the pharmaceuticals

Table 7.1 Innovative conditions and barriers in German biotechnology

	All firms' views (%)		
	True	Partly true	Untrue
Technology gap in the pre-BioRegio mid-90s?	79	21	0
basic research	16	28	56
applied research	45	48	6
commercialization	85	15	0
Improvement post-BioRegio?	36	55	9
Biotech Innovation Barriers			
over-regulation	36	48	15
lack of entrepreneurship	27	48	24
lack of social acceptance	27	55	18
qualified researchers	6	21	73
technology-transfer	50	44	6
inter-firm cooperation	19	63	19
low regional interaction	36	55	9
Lack of public funding	15	42	42

Source: Based on Dohse (1999)

Table 7.2 Firms' assessment of BioRegio contest

	All firms		Non-participants		Non-recipients		Recipients	
	Yes	*No*	*Yes*	*No*	*Yes*	*No*	*Yes*	*No*
continue with BioRegio	75	25	88	13	58	42	83	17
German biotech more competitive	91	9	75	25	100	0	92	8
new jobs created	72	28	63	38	67	33	83	17
more venture capital available	84	16	75	25	100	0	75	25
BioRegio reaching leading innovators	43	57	17	83	27	73	73	27
regional competitions good	58	42	43	57	50	50	75	25

Source: Dohse (1999)

industry by better intellectual property rights (IPR) protection and receiving even more public funding, thus strengthening their bargaining position for early stage product development concepts. Another view offered was that future support should be more for emergent or potential rather than already existing regional biotechnology clusters. As we have seen, the prospect of major public second round funding is more likely to depend on the extent to which venture capital funds can grow to the required scale rather than massively increasing public budgets for biotechnology support. The pharmaceuticals giants remain the most suitable alternative financiers of start-ups at later stages of development, despite the likelihood of further growth in venture capital. Public funding will remain in demand, as everywhere, for the high risk, relatively small sums needed to move a laboratory discovery through to 'proof of concept' stage where it looks possible that a marketable product might emerge after the regulatory three rounds or more of trialling. In Germany, such funds exist as start-up support schemes such as the BioProfile and BioChance initiatives.

As well as assessments from firms, the views of a range of academic, consultant and policy experts on the impact of the policy were elicited. Again there was widespread agreement that before its implementation in 1997, two years after the launch of the contest in 1995, a German biotechnology industry barely existed. Hence the contest had boosted the industry not only in the three winning regions but also in the other fourteen competing regions. Further, because of international awareness of BioRegio, Germany became a focus of interest for international venture capitalists. There was evidence in all three winning regions of some biotechnology firms moving in to take advantage of the supportive hard and soft infrastructures, ranging from premises to consultancy, seed capital and industrial liaison that BioRegios have in place, with access to BioRegio grants a further incentive. Venture capital increased from a paltry DM75 million in 1996 to a still comparatively modest DM425 million in 1998. BioRegio was seen as having contributed to this increase to a considerable extent, though the sums available remain comparatively small. The number of biotechnology start-up firms grew in the same period from twenty-eight to fifty-nine, of which seventeen were in Rhine-Neckar, eighteen in Munich and eleven in Cologne–Düsseldorf. Some firms moved their operations into BioRegios from outside, highlighting the clear trade-off between BioRegio's function in strengthening already strong economic regions and federal regional development policy which seeks to strengthen those that are less favoured. As was argued from the start in this book by reference to the imbalances inherent in the disequilibrium that characterizes disruptive economic change, these are the kind of paradoxical effects that beset policy-making to build clusters. Clusters are inherently disequilibrating economic phenomena and it is the task of policy both to facilitate their benign evolution and seek ways to compensate unfavoured regions for their disadvantage while supporting growth there simultaneously. Despite the relative sophistication of the German cluster-building strategy for biotechnology, it cannot be said unequivocally that this paradox has been satisfactorily resolved. This policy conflict led to calls for biotechnology support to be targeted away from successful clusters towards encouraging developments elsewhere. It was further fuelled by a sense that BioRegio discriminated against innovative firms outside the seventeen regions participating in the contest since they were unable to access funding in the way BioRegio participant firms could.

BioRegio fits into a lengthy history of German federal and *Land* schemes to support biotechnology but differs from its predecessors by its success, relatively speaking, in giving a stimulus to the commercialization aim which has often been the ambition of previous programmes, but never satisfactorily fulfilled. Other programmes complement BioRegio and it is probable that public funding, modelled on this initiative in the sense that it supports start-up companies, will continue though whether with a regionalized rather than national focus is open to question (BioProfile, the 1999 initiative retains this overtly, BioChance does so less overtly). It is thought by critics that the present approach of backing rather than picking winners has been unfair to innovative firms outside BioRegio areas, but the geographical focus and development of firms in proximity to research institutes and local venture capital mean that whatever policy measure is adopted to boost start-ups and support second and subsequent round funding, clusters will remain the leading mode of business organization in biotechnology in future as they have in the active biotechnology economies of the US and the UK This is because clusters offer 'dynamic' external economies or 'spillovers' (Zucker *et al.*, 1998) that promote productivity, innovation and new business formation. This is what drives the competitiveness of biotechnology firms compared to slower-moving 'big pharma', which, in Germany, has been heavily dependent on foreign biotechnology firms to enter the market. Public funding is extraordinarily important to the German venture capital industry and it is hard to see withdrawal of government intervention but it cannot remain a central part of the future of German biotechnology industry development. German biotechnology firms and independent observers agree that BioRegio helped close Germany's technology commercialization gap in biotechnology, and that in some version BioRegio should be continued in future, when the five-year funding period ended in 2002. BioProfile and BioChance continued this trajectory from 1999 onwards. Weaknesses remain in Germany's biopharmaceutical industry, and it is conceivable that a new funding regime will be put in place to help tackle the bias towards 'platform technologies' and away from therapeutics.

How does the cluster-building process function locally?

The German federal government took an interesting and innovative step by recognizing previous failures ensuing from a narrowly technological focus and opting instead for a regional clustering approach as the vehicle for an industry policy designed to close a widely perceived technology gap and transform a dormant sector into one intended to be globally competitive. This is one of the most pronounced policy shifts in advanced economy industry policy revealing policy learning that clustering in biotechnology is perceived as advantageous and widely practised in the USA and the UK, something which BioRegio sought to emulate. In what follows, accounts of the status, specialities and particular strengths of the three winning BioRegios (plus Jena) are provided and some estimate given of the nature and degree of clustering activity associated with them. Data sources used have been of a largely secondary nature, supplemented by some personal communication with key actors or academic experts in each of the three main BioRegios. It will be recalled that the three key BioRegios are: Rhineland, comprising Cologne, Düsseldorf, Wuppertal and Aachen; Rhine-Neckar, including Heidelberg, Mannheim and Ludwigshafen, and Munich.

Rhineland BioRegio

Given its history as a heavy industrial region undergoing major restructuring away from coal, steel, chemicals and heavy engineering towards newer growth industries, the land of North Rhine Westphalia has launched numerous technology-orientated initiatives, especially from 1984, when a number of new technological institutes with near-market research functions, technology parks and innovation networks were set up under the TPZ (Future Technology Programme) initiative. Among these was the land's first biotechnology programme (*Landesinitiative Bio- und Gentechnik e.V*) to support small and medium-sized enterprises. Other sectors receiving support included environmental technology, energy technology, micro-electronics, measurement, IT and materials. The biotechnology initiative was superseded in 1994 by the establishment of BioGenTec. This agency is a non-profit organization with representation from industry, academia, trade unions and government. It acts as an intermediary body linking biotechnology start-ups, an expert network of 200 members, venture capitalists and partners from industry. It aims to become a commercial company and will sell services to the industry on that basis. A wide range of medical biotechnology areas is prioritized under its programme of support, and environmental and agro-food technologies are also promoted.

Various networks have been established, including a venture capital network of fifteen local and internationally operational firms and groups, a competence and training network, and a management and coaching network. BioGenTec has established Biocentres at various locations and organizes an annual international meeting called the BioGenTec Forum. In 2000 an international forum on nanobiotechnology was organized. The research strengths of the land include the Max Planck Institute for Plant Breeding research at Cologne, which has become the centre of agro-food biotechnology, around which larger (e.g. Monsanto, DSV and Agrevo) and smaller firms are clustered. In 1998 a letter of intent was signed between the governments of North Rhine-Westphalia (NRW) and Saskatchewan, Canada, to improve collaboration in the field of agro-food biotechnology. Also in Cologne is the Max Delbrück Laboratory (another part of the Max Planck Society) specializing in plant genetics. The Max Planck Institute for Neurological Research, specializing in (photo)receptors, signal transduction and recombinant proteins is at Mülheim an der Ruhr. A Helmholtz Institute exists at Aachen (Biomedicine and Cryobiology), and a Fraunhofer Institute for Environmental Chemistry and Ecotoxicology at Schmallenberg. Altogether the *Land* has some 167 research institutes, many employing relatively small numbers of researchers, but with representation across the biopharmaceutical, agro-food and environmental biotechnology spectrum.

In respect of relationships between multiple governance levels, *Land* and federal support programmes fit together well, as the transition from TPZ to BioGenTec suggests and it is clear that they have been complementary sources of funding over the years. The division of funding responsibilities has been one in which federal funding went to expanding companies while *Land* funding went directly into start-ups. This led to a steady movement of companies into the BioRegio area so BioGenTec established regional offices outside the BioRegio area (Münster and Bergkamen, north and east of the Ruhr) to seek to seed new clusters by attracting start-ups to establish in these more outlying areas. In terms of the sectoral distribution of biotechnology firms 22 per cent are in diagnostics, 12 per cent pharmarceuticals, 7 per cent agro-food, 18 per cent in environmental protection, 9 per cent filtration engineering and 10 per cent bioanalysis. The last three are primarily engaged in environmental biotechnology, making that the largest category. This is seen as a 'Cinderella' part of biotechnology and one in which it is hard to get university-derived start-ups underway. However, it is a strength of this region's biotechnology profile mainly due to its origins in the regional restructuring efforts of the NRW government. German co-determination rules require union consultation on issues like redundancies and downsizing. Hence management could not simply decide to close down redundant plants but were required to explore alternative trajectories firms might seek to move along. Because of expertise in the mining and steel industries of filtration and ventilation technologies, it was recognized that adapting these for environmental clean-up could be promising. Diversification by larger firms and establishment of independent corporate spin-offs occurred in the 1980s in the Ruhr to create a supply sector for the 'greening' of the landscape. Fortuitously, this industry was then ready for the large task of effecting reclamation and clean-up in former East Germany and thereafter Central and Eastern Europe where there was growing demand for such services. This also anticipated tougher new environmental regulations being framed in Germany, and later the EU. In any case, between 1984 and 1994 some 600 firms turned towards environmental technologies, a small portion of which applied biotechnologies. Some 100,000 jobs exist in this regional environmental technology industry, which itself has been shown to have a cluster-like character (Rehfeld, 1995).

The question remains as to whether Rhineland biotechnology sector has taken on a cluster character as a consequence of BioRegio. Many key infrastructural conditions are present including a strong science base, expanding numbers of firms, qualified staff, advanced laboratory space such as the Rechtsrheinisches Technologie Zentrum (RTZ) in Cologne, a 5,000-metre square biotechnology incubator with a C4 (high quality clean room) central laboratory, availability of finance, business support services, a skilled workforce, effective networks and a supportive policy environment. However, the number of biotechnology start-ups was at a peak of twenty-six in 1990–91, and in decline to a figure of twelve in 1994–95 with only eleven start-ups and eight company expansions from 1996 to 1999. This is consistent with the cautious approach, on the one hand and the difficulty of setting up new firms in environmental biotechnology, on the other. For example, even in the leading joint biotechnology centre of Aachen and Jülich start-ups have been hard to stimulate. So the Rheinland BioRegio has all the appropriate conditions for stimulating the development of reasonably large numbers of new biotechnology firms, but whether a significant cluster of growing biotechnology firms will appear swiftly must remain doubtful on the evidence presently available.

Rhine-Neckar-Dreieck

Heidelberg is Germany's oldest university and has one of the best science bases for biotechnology. Two Max Planck Institutes, for Cell Biology and Medical Research, are in the region, as is the German (Helmholtz) Cancer Research Centre (DKFZ). Located there also are the European Molecular Biology Laboratory, the European Molecular Biology Organization and one of the four Germany's Gene Centres, the product of an earlier round of federal funding to stimulate commercialization of biotechnology. Also present are the Resource Centre of the German Human Genome Project, two further medical genetics institutes and two plant genetics centres. Three other universities, Mannheim, Ludwigshafen and Kaiserslautern and three polytechnics complete the research and training spectrum. A number of Germany's leading pharmaceuticals firms are part of the regional cluster area, including BASF/Knoll (Ludwigshafen), Boehringer Mannheim Roche Diagnostics (Mannheim), and Merck (Darmstadt). But the heart of the BioRegio is the Heidelberg-based commercialization organization, the Biotechnology Centre Heidelberg (BTH). This is a three-tiered organization consisting of a commercial business consultancy, a seed capital fund and a non-profit biotechnology liaison and advisory service. Central to BTH's functioning is Heidelberg Innovation GmbH (HI), a commercial consultancy that takes company equity in exchange for drawing up market analyses, business and financial plans, assisting in capital acquisition and providing early phase business support for start-ups. It is a network organization, relaying information, partnering organizations seeking contact with local biotechnology companies and linking to research institutes and local authorities.

The key initial financing element of BTH is BioScience Venture. This financial vehicle was founded by local pharmaceuticals companies and banks. It is managed by HI and acts as a seed fund and lead investor in early start-ups. It also seeks international venture capital to finance second round developments. Assessments of project viability are made with advice from HI and BioRegio Rhine-Neckar e.V., the third element of BTH. The last-named seeks out commercial projects and recommends the most promising for BioRegio public funding support. Business proposals have run at some fifty per year since 1996, but between 1996 and 1998 only nine start-ups had been established, a figure that had risen to seventeen (including biochip and biosoftware firms) by 1999. The total number of biotechnology SMEs (excluding start-ups) was twenty in 1998. Most are in the healthcare sector, with some in plant genetics. The main location for this cluster of some thirty-seven biotechnology firms is the Heidelberg Technology Park and the adjoining Biopark on the university's science campus. This has 10,000 square metres of laboratory and office space plus a further 6,000 on the Production Park nearby, to where start-ups move once they have grown beyond the research phase. A joint venture by local firms and universities has been to establish the Postgraduate BioBusiness Programme. This is designed to provide scientists with hands-on experience of business administration through three months' coursework and nine months of practical training in industry.

Once more, key ingredients for successful clustering are present, including close proximity for firms on the technology park to 'big pharma' in Ludwigshafen and Mannheim and leading-edge science in Heidelberg. The *Land* of Baden-Württemberg has a biotechnology initiative but also distributes its funding among the Freiburg BioValley (one of Germany's most dynamic biotechnology locations), Ulm, and Tübingen-Stuttgart as well as the Rhine-Neckar region. BioRegio funding is principally used for start-ups, most of whom are currently making losses. But through the network-like character of BTH, lead investor capital from BioScience Venture can be tripled by leveraging both BioRegio funding and *Land*/corporate venturing funds. Thus reasonable sums of start-up capital can very easily be raised at low risk to the lead investor. The *Land* helped fund Heidelberg Technology Park, subsidizes a patenting support initiative, providing grants to universities for making patent applications, and funds a Young Innovators pre-start-up funding programme for university and research institute personnel.

Some examples of firm practices and specialities give a flavour of activities in this BioRegio. Three cases show a variety of firm origins: the big pharma partnership, the research institute start-up, and the 'rent-seeker' entrepreneur. The rent-seeker is the founder of Molecular Machines and Industries (MMI), former holder of a chair in biochemistry at Regensburg near Munich, but doctoral student at Heidelberg where he established his first start-up SL Microtest in 1993. This company (optical cell tweezers) has grown into a well-known optical bioinstrumental firm. Funding was self-generated but, to grow, the firm moved to Jena (former East Germany) where a more favourable grant-regime operated and optics is a famous local speciality. There a new fluorescence screening device was developed. HI was approached and advised on a business plan, contracts were signed in early 1998 and space found in the Heidelberg Biopark. A different firm, Lion Bioscience, an abbreviation of Laboratories for the Investigation of Nucleotide Sequences, aims to be a European leader in genomics and bioinformatics. Funding of DM4.5 million was raised privately by six Heidelberg scientists, and an automotive industry entrepreneur. BioRegio co-funded with DM2.5 million. Technology was transferred from the European Molecular Biology Laboratory and the German Cancer Research Centre in Heidelberg, as well as the European Bioinformatics Institute at Cambridge. The company's key bioinformatics tool is bioSCOUT a powerful bioinformatics management system. Established in 1997, the firm had 100 employees by 1999. Finally, BASF-LYNX AG is a German-US joint venture, LYNX being a therapeutic treatment firm. BioRegio funding facilitated the partnership, an overall company investment of DM100 million secured the joint venture. Activities include treatment of epilepsy in collaboration with the local Max Planck Institute for Medical Research, in receipt of a BioRegio grant, screening for toxicity detection, and development of micro-organisms for production of amino acids and vitamins. When start-up funds are exhausted, an IPO may be sought, or continued third-party collaboration.

It is less difficult to see focused cluster potential in the Heidelberg region than in Cologne, not least because it has strength in depth scientifically and organizationally in biopharmaceuticals. The involvement of local pharmaceutical multinationals means that important alternative financing of product development can be envisioned. The key problem presently is the immaturity, small-scale and platform technology emphasis of the start-ups. The prospects for collaboration are lower here because the sector is rather cut-throat. Alternatively, many now leading biotechnology firms started off in diagnostics and the like, progressing to become therapeutic drug producers, where high profitability lies, afterwards. Currently, therefore this is too small and too vertically linked to individual interactions between entrepreneurs and the science base to be considered a functioning cluster. Whether or not that is feasible will depend on the risk investment profiles that emerge at second and third-round funding stages.

Munich

The commercial application of biotechnology is claimed by the local industry to have begun in the 1950s when Boehringer Mannheim moved part of its diagnostics R&D to Munich. More recently this company invested DM150 million in production facilities in a southern part of Munich. But Martinsried and Grosshadern in the south western suburbs mark the centre of Munich's biotechnology. Hoechst Marion Roussel (now Aventis) opened its Centre for Applied Genomic Research there and the Biotechnology Innovation Centre (IZB), funded by DM40 million from the Bavarian government, is located nearby with 9,000 square metres of laboratory and office space. The organization responsible for managing development of biotechnology, BioM, is also located at Martinsried. The area has become a biomedical research campus with 8,000 researchers working in biology, medicine, chemistry and pharmacy located there. Unlike Rhine-Neckar's BTH, BioM AG is a one-stop shop with seed financing, administration of BioRegio awards and enterprise support under one roof. Seed financing is a partnership fund from the Bavarian State government, industry and banks up to DM300K per company. BioM's investments are tripled by finance from TBG, a public reconstruction bank and Bayern Kapital, a special Bavarian financing initiative. The latter supplies equity capital as co-investments. The fund has DM80 million for supporting biotechnology activities. Bay BG, and BV Bank-Corange-ING Barings Bank have special public/private co-funding pools, and a further eight (from sixteen) Munich venture capitalists in the private-market sector invest in biotechnology. By 1999, sixteen start-ups had been funded to the tune of DM59 million with a third of this coming from BioRegio sources. BioM is a network organization, reliant on science, finance and industry expertise for its support committees. It also runs young entrepreneur initiatives, including development of business ideas into business plans and financial engineering plans. Business plan competitions are also run in biotechnology.

The science base in Munich is broad, but with special expertise in health-related and agricultural and food biotechnology. There are three Max Planck Institutes (MPIs) of relevance, in Biochemistry, Psychiatry and the MPI Patent Agency. GSF is the Helmholtz Research Centre for Environment and Health, and the German Research Institute for Food Chemistry is a Leibniz Institute. There are three Fraunhofer Institutes, one of Germany's four Gene Centres, two universities and two polytechnics. The main research-oriented big pharma companies are the aforementioned Roche Diagnostics (formerly Boerhinger Mannheim) and Aventis. The work areas of this science community include: 3D structural analysis, biosensors, genomics, proteomics, combinatorial chemistry, gene transfer technologies, vaccines, bioinformatics, genetic engineering, DNA methods, primary and cell cultures, microrganisms, proteins, enzymes and gene mapping. The Bavarian commitment to biotechnology (and other new technologies) was realized in consequence of a *Land* government decision to part-privatize shares in regional power-generation and distribution companies in the 1990s, thereby creating a funding pool to subsidize applied technology developments. Such commitment was further expressed in exercising *Land* rights to grant permissions for genetic engineering earlier and with fewer obstacles than in the other German *Länder* once federal legislation had been passed. Such permissions are *Land* rather than federal responsibilities, The Bavarian Ministry of Economics adopted a local version of the US model of commercialization on the consultancy advice of the Fraunhofer Institute for Systems Innovation at Karlsruhe – venture capital, management support and start-ups to transfer research results from laboratory to market. As we have seen elsewhere, though, this is mostly sought through public initiative with private venture capital joining as co-financiers.

In common with the other BioRegio winners, Munich's vertical networks from science through (public) funding to start-up are, in principle, strong, though, as elsewhere, given the almost risk-free funding regime, the numbers of start-ups are not overwhelming, again due to the quest for 'quality' start-ups in whom substantial sums may be individually invested. A further explanation for conservatism is that BioM AG is set up as a corporation and makes investments with its shareholders' (state, industry and banks) money. Most of the shares are held by banks seeking to earn high returns. The banks also use this method to learn about biotechnology, its risks and prospects. Thus, an already well-subsidized system is further protected from risk by the influence of banking culture, itself highly conservative in Germany, to ensure, as far as possible, risk avoidance. Hence, while BioM is the network face of the biotechnology cluster in Munich, its activities are ultimately orchestrated indirectly and directly by the banks, abetted by a fairly risk-averse, mostly publicly-funded, venture capital industry and the local pharmaceutical and chemical companies (see Giesecke, 1999).

With respect to *Land* and federal relationships on funding, Munich BioRegio, once again, demonstrates the seamlessness of the fit between programmes. This is no surprise since a great deal of 'concertation' proceeds between the two levels of government on a constant basis and the last thing either wants is a resort to the Constitutional Court to rule on inter-governmental conflicts. Hence, this is a further illustrative instance of the German consensus-oriented mode of policy-evolution. Similarly, the consistency with which public, scientific and industrial partnership characterizes funding or technology-transfer mechanisms is illustrative of the ingrained networking culture that characterizes German governance. As to whether Martinsried and Munich more widely constitute a cluster, the answer is probably positive although there are conflicting reports as to whether three key firms commercializing biotechnology from Max Planck Institutes are interacting, collaborating companies, or not. Dohse (1999) suggests that despite their common origin they are not strongly linked. But Clarke (1998) notes that two of them, MorphoSys and Micromet are collaborating on the development of an antibody-based treatment for micrometastatic cancer. MorphoSys was the first firm to receive a BioRegio grant and had previously collaborated successfully with Boehringer-Mannheim on the development of a diagnostic reagent. MorphoSys' business strategy is to focus on the development of horizontal networking. The firm has no plans to develop therapeutics, aiming to remain a science discovery firm and let partners carry the risk of drug development. Thus MorphoSys works with a variety of companies, minimizing its risk-profile, but potentially benefiting from substantial injections of capital from research funding, milestone payments from 'big pharma' and royalties. MediGene is another Munich company that, unlike MorphoSys, does plan to become a fully integrated biopharmaceutical company. It was a spin-out from Gene Centre in 1994 and has raised DM23 million from the typical German sources: venture capital, state and federal funds. Its expertise is in gene therapy for cancer and cardiovascular diseases. MediGene has alliances with Aventis on gene therapy vectors and a vaccine for malignant melanoma. Academic–clinical partnerships include the Munich Gene Centre, the Munich University Hospital, the German Cancer Centre at Heidelberg and, in the USA, the National Institutes of Health and Princeton University. Its co-founder, Horst Domdey recently gave up a chair at Munich University to become head of BioM. Mondogen spun out of the Virus Research department at the Martinsried Max Planck Institute for Biochemistry. Its founder was director of the department and had co-founded Biogen, one of Boston's oldest biotechnology firms, in 1978. Advice from a local incubator and technology park, IZB and funding from BioRegio plus victory in a McKinsey Business Plan competition led to the founding of Mondogen. Martinsried is said by Mondogen's founder to be unlike MIT or Cambridge as a cluster, but to have the 'seed crystal' for such a future. The main obstacles are the different cultures between German scientists and venture capital 'speculators', the one profoundly committed to science for its own sake, the other deaf to such high-mindedness.

Jena

This is the city in former East Germany given a special prize in the BioRegio contest. Its co-ordination office is BioRegio Jena e.V., established in 1997 and forming a network consisting of an incubator (forty firms) part-specialized in medical and environmental biotechnology, an economic development agency – BioStart mbH, and a venture capital organization Thüringer Wagnis-Kapital Fonds. Optoelectronics is Jena's historic industrial specialization (said by Hendry and Brown, 1998 to be a functioning cluster) and, along with precision engineering, makes for distinctive links with biotechnology. There are Max Planck Institutes for Chemical Ecology and Biogeochemistry, a Leibniz Institute for Molecular Biotechnology and institutes for Natural Products, Bioprocess Engineering, Medical Engineering and Biotechnology research. Two universities and a polytechnic complete the scene. A joint Department of Trade and Industry and Foreign and Commonwealth Office (1998) report notes twenty or more biotechnology SMEs including fourteen start-ups since 1996. These form 'a close-knit science community with extensive collabora-tions' (König, 1998). One such collaboration between the Institute for Molecular Biotechnology and the Freidrich Schiller University led to advances in understanding 'Sak', creating rSak and in collaboration with the Natural Products Institute, reaching Phase III clinical trials for GMP certified production of this treatment for re-opening arteries after acute myocardial infarction. Opal Jena is a spin-out from Jenoptik (formerly Carl Zeiss) in biotechnology instruments, with a 5 per cent world market share for its Plate Mate automated pipettor. Food Jena is a biotechnology company focused on the development of functional foods (i.e. with therapeutic benefit). It is an academic start-up from an agro-environmental science background, working entirely on animal foodstuffs. Jena benefits substantially from its high quality human capital. and Clondiag Chip Technologies is a PhD student start-up providing substance libraries on a chip, having negotiated seed capital of DM5 million, one-third from an Austrian corporate partner, Electronic Visions.

The winning BioRegios each have exceptionally strong enterprise support infrastructures complementing strong local science bases. Network links between actors are pronounced, with co-funding of venture capital between public and private sectors the norm. There are difficulties in getting large numbers of new businesses up and running despite the apparently generous grant-aid available. This seems partly explicable by the risk-aversity of the lead investors and the conservatism of the banks that are often closely involved behind the scenes in the management of BioRegio cluster development. In all cases *Land* and federal funding regimes co-exist happily, and in some cases, initiatives set up by the lower level of government are easily absorbed into new initiatives, notably BioRegio, emanating from the federal level. Perhaps one of the most striking features of the government, industry and science relationship in respect of biotechnology is quite how inter-woven they are into what Etkowitz and Leydesdorff (1997) call 'the triple helix', even at the city level of operations. Communication levels between key actors are high quality with no clear evidence of withholding information, networks function effectively and 'seed crystals' for possible future clusters have been sown in numerous regions of Germany. The really testing time for biotechnology firms, whether in latent clusters or not, will come when large doses of second round funding are needed as firms move towards therapeutic drug-production. This will begin occurring seriously around the year 2002.

Conclusion

This chapter was written with a view to extracting policy lessons from instances where governments at various levels have sought to assist in the building of clusters through designing interventions meant to overcome or respond to market failures. The chapter looked at examples of decaying, declining and possibly reviving clusters drawn from the Midlands and North of England, the country where they first emerged.

It then examined an extreme case in Finland set in the wider context of cluster-building policies in Nordic countries where success had been achieved in a geographically remote, declining industrial setting. Finally, a detailed account of one of the most ambitious regional cluster-building strategies from Germany was given. What can be learned from these examples? Three main lessons can be drawn to assist in learning clusters, and the lessons come as vividly from the failures as from the successes, rather as Orsenigo's (2000) paper on the rare subject of failed cluster emergence in Italy does. But the simple lesson is that cut-throat competition works against competitive, let alone co-operative advantage, while vertical interaction in the governance sphere and horizontal interaction and regional partnership for collective learning produce promising results for policies seeking to promote Knowledge Economies.

First, failing clusters have in common a characteristic of stubborn resistance to change, a cavalier collective attitude towards the value of knowledge other than that traditionally available, and a culture promoting the withholding of information of central importance to normal, day-to-day business functions. From the perspective of this book that adds up to a perverse, irrational way of doing business. But it may once not have been so. Change is only welcomed when stability is no longer an option, and many firms in older clusters in the UK grew in an era of reasonable economic stability under imperial preference and with the vestiges of first-mover advantage. Individualism implies capability to maintain confidences, particularly those germane to the specific technical or commercial knowledge discovered that gave firms some first-mover edge. And even not investing in market or business cycle intelligence, if by so doing an expense is incurred which competitors will not have to pay but may learn on the grapevine, can look rational viewed from a particularly obtuse angle. But it all represents a 'tragedy of the commons', an obvious case, as in game theoretic classics such as the Prisoner's Dilemma (Axelrod, 1984) where collaboration to some degree, the furthest admissible, has to be superior to head in the sand, individualistic competition. Successful clusters and aspiring ones will contain actors who are more open, more willing to exchange ideas, information, even knowledge, and engage in trustful interaction than failing ones. This is not easy, as the details of cultural obstacles between scientists and corporate venturers and venture capitalists show in Germany, but there are 'systems in place' in the form of localized cross-functional networks that are there to help overcome such obstacles.

Second, unsuccessful clusters operate under more stressful conditions than successful ones, partly because of the previous point, partly because they find themselves in survival, cost-cutting mode under circumstances where there is not an abundance of either knowledge capital or financial capital, let alone social capital, that can act as the means of enabling time for strategic thinking and action to be bought. Crisis management is said to force decisions but the array of possible choices is by then so small that the decision taken will inevitably be non-strategic and sub-optimal. Cluster relations still operate under such conditions but by way of a systemic downward spiral which causes repercussions up and down supply chains as was shown in the case of Lucas, a key Rover supplier that was in the 1990s carved at least four ways by US competitors, first by Computer Sciences Corp., then by Varity and Caterpillar, finally by

TRW because of its Byzantine organizational structure and inability to learn the virtues of collective learning. This, of course was not helped by the dissimulations and worse of its key automotive customer, Rover, the vicissitudes of which had nearly proved fatal to the Birmingham automotive cluster.

Third, successful modern clusters thrive on a high valuation of scientific knowledge whether about discoveries in their field of competence or economic assessments of the trajectory of their industry. Yet they are not merely data-driven entities. An important element is a kind of community recognition of commonalities and possible complementarities among diverse members of potential or actual cluster. The existence of image, identity and vision and the ability to transform that into a consensus for action along specific, collectively beneficial lines is important. So are the sharing of responsibilities for leadership and ensuring that collective requirements are not only listened to, but also acted upon. This is pronounced in the German cluster-building policy for biotechnology, perhaps too much so, since as yet the outcomes are limited. But establishing leadership and organizational structures that can deliver key resources previously absent but necessary for cluster-based economic development is also vital for policy to have a reasonable chance of success. One danger from close observation of the German approach is conservatism of a generic kind. US venture capitalists are specifically conservative when it comes to early stage, high-risk funding but major risk-takers once they see a chance of success. They can see the latter quickly because they are intimately acquainted with the scientific as well as the financial engineering requirements of knowledge-based entrepreneurship, something which, as was shown, German banks were hoping to 'learn by doing' through their co-funding investments.

8 Knowledge Economies
Here to stay? Where to go?

Introduction

Before starting a reprise of the book's main argument in the following section, it is worth laying out grounds to help draw conclusions about the likely longevity of the more general phenomenon of 'knowledge-based economies' and the particular forms it may evolve into. One of the things that can be done in justification of the premise on which the book's themes arise is to mobilize certain kinds of macro-economic data that validate an argument. Had this book been about the generic phenomenon of the growth of a knowledge-based economy, indicators such as those about to be alluded to would have had a more central role. But this book has been about Knowledge Economies, meaning localized and regionalized, clustered, collective learning systems. In themselves these are relatively rare phenomena, though where they have come into existence in the USA, this has occurred rapidly as we saw in the cases of Richardson, Boston and as Saxenian (1994) and others show for Silicon Valley. The last named now has some 6,000 firms when in the 1970s it had less than 600. They also show resilience.

This is being written as the London and New York Stock Exchanges are being officially described as having 'crashed', i.e. lost 20 per cent of their value in a year, a factor brought on in large measure by the volatility of New Economy stocks. Starting with the obviously over-valued dot.coms, the contagion spread from Internet content providers to the kings of the superhighway, Cisco, Lucent and Nortel, then to the telecommunications service providers, then to the software and microprocessor producers. But in San Mateo, at the northern end of Silicon Valley, e-businesses specializing in contract management software have sprung up, close to leading corporate software firms like Siebel Systems and Oracle, but expert at designing software to enable small firms to network more effectively and efficiently in or out of clusters (Kehoe, 2001). What is particularly interesting about this is that in a paper to a seminar on Cluster Policy in Nordregio, Stockholm, Maria Chiarvesio from Venice International University reported that 'Enterprise Resource Planning' (ERP) firms such as Oracle, Lotus and German competitor SAP, had set up offices in Venice to study why firms in nearby industrial districts were low users of e-commerce, study their modes of interactive networking for the full range of business activities, and develop inter-organizational software that would mimic for clusters what ERP did for the internal communications efficiencies of large corporate organizations (Chiarvesio, 1999). Developments in San Mateo in early 2001 suggest that learning about cooperative advantage occurred and that spin-off or connected firms are responsible for knowledge development and commercialization. Kehoe reports that Siebel was formed by a former Oracle executive, and AserA, a new business (also one of the Kleiner Perkins *keiretsu* or cluster firms) has several former Oracle senior staff.

This is analytically important because it signifies the ascendancy of the small business to recognition by large corporate software suppliers like Oracle realizing that their corporate market has become saturated, for the time being, with management software solutions. Apart from public organizations like government departments that have often learned expensive lessons about the low functionality for them of such products, small and medium-sized enterprises are the only remaining market. But it is even more important intellectually and for the theoretical project of this book because it encapsulates all its master concepts while opening a new door on key business processes of multinational businesses, at least in a core Knowledge Economy industry like software.

For, hitherto it has always tended to be a dominant view in economic geography and regional science that spatial development occurred either, in the early Marshallian or even Weberian perspective as discussed in Chapter 2, by interactions in a specialist industrial district where innovation was 'in the air', or more recently, particularly after the work of Vernon, also discussed in Chapter 2, through the global supply chain. In both cases, to use the insights of theorists of development in less developed economies like Gereffi (1996; 1999), Altenburg and Meyer-Stamer (1999), and Humphrey and Schmitz (2000), the 'governance of upgrading' is the key to development. In clusters, this governance is, as we have seen, fundamental to maintaining upward evolutionary trajectories, otherwise the cluster atrophies. It may be market-based, or a hybrid, associative forum which facilitates public–private interaction. There will be multi-level dimensions to this governance, as we have seen.

Alternatively, the governance of upgrading, meaning greater efficiency, quality and innovation in production, is conducted at the behest of customer or producer multinationals in the global supply chain, forcing suppliers to improve and enabling them, in principle, to be more globally competitive.

But in Knowledge Economies, while both impulses in the governance of upgrading are undoubtedly pronounced, there are crucial exceptions. First, as the Cambridge (UK) instance, not to mention biotechnology clusters practically everywhere, showed, directionality is reversed since large corporations themselves seek upgrading, in the sense of research and innovation knowledge, from the clusters. Many large corporations have closed or attenuated their R&D functions and routinely buy from the research and innovation supply chain. Second, governance by buyers or producers in global supply chains may not be as total as those who postulate it imagine. This is because of the need for local solutions to problems created by application of advanced knowledge in, say, speeding up time-to-market, for less developed economy suppliers who do not have the absorptive capacity to meet what to them may be unrealistic demands. Hence, the upgrading process is increasingly be governed by the 'slowest ship in the convoy'.

But, for our purposes, the implication of the 'contract management software' example is that it represents knowledge acting upon knowledge itself to create productivity not through cluster or supply chain governance, but by cluster-to-cluster governance. This can sound newer than in reality it is, for it is known that knowledge transfer through market transactions, in the main, between, for example, ceramics firms in Sassuolo in Emilia-Romagna and Castellon in Valencia are long established, and the latter have a similar relationship with the ceramics cluster in Santa Catarina, Brazil (Altenburg and Meyer-Stamer, 1999). However, the key point is that as clusters become more and more the repositories of key, advanced knowledge as they are in the Knowledge Economies, not least because larger firms conduct less leading edge research than they used to, more of this cluster-to-cluster learning is occurring through and from knowledge of cooperative advantage in enhancing innovation and competitiveness. It is pronounced in biotechnology as the discussion of international linkage in research and marketing among Cambridge, Boston, Oxford and San Diego clusters in Chapter 6 showed but further research is needed to assess how normal it has become among New Economy clusters.

The undulating landscape of the Knowledge Economy

The book is about Knowledge Economies but these are peaks among undulating foothills, plains and valleys of relative Knowledge Economy deprivation. First, the general background that underpins the thesis of the book, something echoed sectorally by the discussion of the Norwegian food industry in Chapter 1 is important. Dunning (2000) shows that, whereas in the 1950s, 80 per cent of value added in US manufacturing was raw materials and 20 per cent knowledge, by 1995 these proportions had reversed to 30 per cent and 70 per cent respectively. Moreover, the book value of US corporate *tangible* assets was estimated to have declined from 25 per cent–33 per cent in the 1980s to between 5 per cent and 20 per cent in the late 1990s. Throughout the OECD area, from 1975 to 1995, R&D expenditure rose three times as fast as manufacturing output, patents registered in the USA rose by 48 per cent and those in knowledge-intensive sectors by 182 per cent, and the student cohort share rose from 35 per cent in 1980 to 56 per cent in 1993. All these indicators point inescapably to the rise of the knowledge-based economy.

In a study of where the OECD-defined knowledge-based industries are located in the UK, Cooke *et al.* (2001a) found four regions to be above average and seven well below, using 1998 data. That imbalance was because of London's and southern England's predominance, ranging from 50 per cent above the UK mean in the case of the capital city, to 27 per cent above for south-east and south-west England together. The Midlands and North of

England hovered at some 75 per cent of the norm, Northern Ireland, Scotland and Wales either side of 70 per cent. Returning to London, even its deprived north and eastern parts scored above the UK average, marginally, at 3 per cent whereas islands like Anglesey in Wales (41 per cent) and the Western Isles in Scotland (32 per cent) were low even compared to areas of older industry in decline like Merthyr Tydfil (48 per cent). Thus if the knowledge economy, in general, is to grow in the future as it is likely to do as Old Economy industries migrate to less developed economies, it is clearly advantageous to be a job-seeker in the poorer parts of a metropolis like London than a rural idyll like Anglesey. The forces underlying knowledge-based industry will suck more and more workers into large cities, exactly the opposite effect to that predicted by those who saw the 'Information Age' as harbinger of the death of geography and the 'death of distance' (Cairncross, 1997).

When we look at the localities that have the highest OECD knowledge-based industry index in the UK, they exist like spokes on a wheel, the hub of which is, not London, but Heathrow airport. Bracknell (203 per cent), Wokingham (198 per cent), Maidenhead (166 per cent), Reading (152 per cent) and, a slight outlier, Milton Keynes (150 per cent) are all within easy reach of each other, London, and its principal airport. It is probable that none of these has a specific Knowledge Economy cluster but that they are part of a metropolitan agglomeration of high-technology industry and knowledge-based services at between one and a half and two times the national intensity and more than two and a half times the intensity of the least knowledge-based areas. Other cities in the UK are mild summits in the undulating landscape of knowledge-based industry but small towns outside the Heathrow penumbra and rural areas are the vales. Sternberg (2001) has shown a comparable picture to that of Zook (2000) using Internet concentrations as the key discriminating variable. In Germany, Munich, Berlin, Hamburg and Cologne predominate while in the USA the big four for Internet content and venture capital are San Francisco, Boston, New York and Los Angeles in that order. The UK is different in having so much urban primacy in its capital city region.

In the opening chapter, this book began by offering a variety of ways into an understanding of the gross undulations in social and spatial terms that attend the birth of the New Economy and its specific, predominant form in clusters that were characterized as Knowledge Economies. It defined the latter, in line with Castells' (1996) insight that they are new because they consist in knowledge acting upon knowledge itself for productivity. It is important to remember that Old Economy industries like food processing are, nowadays, notable users of scientific knowledge, itself replacing experience, rules of thumb and more rudimentary science involved before. But final output is food not new knowledge. The food industry is dominated by large, hierarchical corporate processors and supermarket chains that distribute to consumers. These links spread increasingly globally in extended supply chains. In a particular national territory like the UK, even a limited geographical concentration mapping exercise such as DTI (2001) reveals every UK region except London to have a higher than average location quotient in food production of some speciality or other, thus it is a ubiquitous industry. It is probably an exaggeration to say that it is a highly clustered industry in the sense meant here, but it will almost certainly become more so as pressure grows for regulation of the power of corporate control in light of recent Genetically Modified Organism and animal health scares and knowledge intensity of an organizational as well as technological kind becomes more entrenched.

Theoretically, it is straightforward to see why the landscape of clusters is so uneven. It reflects and magnifies the basic characteristic of competitive markets, best understood by Schumpeter and Marx with respect to entrepreneurship and innovation. Both understood the way that 'all that is solid melts into air' or is subject to 'creative destruction' where radical innovation in the economy occurs. Disruptive change is not normal but it expresses 'punctuated evolution' in the broad trajectory of economic events. Authors like Myrdal and Hirschman also understood the fundamental condition of disequilibrium that characterizes capitalist economies. Punctuation points occur when paradigms shift. That occurs when a concatenation of interdependent technologies produces new and more efficient ways of performing crucial economic tasks. Factories that brought together investment, energy, raw materials, labour and innovative organisation of production processes brought disruptive change and creative destruction to craft and homework-based forms of textile production. Electricity undermined steam power and made possible a wide range of new products and services from light bulbs to telecommunications that signed the death warrant for the Pony Express and the oil lamp. Modern chemicals and pharmaceutical industries were unthinkable without electricity disruptively changing life for lime-burners and apothecaries.

These were more pervasively scientific, knowledge-based industries than their predecessors but they were still not creating value from knowledge for further knowledge commoditization. They were not 'light GDP' industries (Leadbeater, 2000) but used science to transform nature into profitable products and services. Even so, as Schumpeter noted, innovations in these times brought swarms of imitators of the novelty to the location of origin though such imitators were quickly swallowed up as ownership concentrated. Nowadays, barriers to market entry are mostly too high for imitation in Knowledge Economies. Even in the dot.com stampede at the turn of the millennium, advantage was mostly sought in identifying niches in a given country's economy though such niches were often mimicked from those in leading Knowledge Economies such as Silicon Valley. Neoclassical economics has struggled to grasp the economics of imbalance, but not very hard until comparatively recently. Krugman's (1995) work is the best of its kind, but ultimately he is self-critical of his own simplistic 'two location' models even though they solve, by relaxing untenable assumptions of constant returns and no uncertainty, problems that escaped the neoclassical masters of traditional regional science. Beyond that, though, the understanding of clusters as Knowledge Economies, their learning and tutoring interactions, and their use of co-operation for global competitiveness, must proceed beyond econometrics and closer to political sociology and anthropology.

In the third chapter a key feature of Knowledge Economies was explored, namely, the governance process involved in interactive learning and innovation. At the heart of the governance question are the issues of legitimacy in determining priorities and the pursuit of those priorities with necessary and appropriate budgets. One of the paradoxes of Knowledge Economies is that the nuances of clusters are best understood at ground level by the actors involved in their institutional and organizational fabric themselves, but often the resources required to allow them to function must come from governments, large charitable trusts or foundations and large corporations. This is the basic research-funding plinth on which clusters rest, even those at the core of the Digital Economy in new media who mostly sell their knowledge-processing expertise to broadcasters, advertisers and publishers to survive.

This is why governance is a key concept in the analysis offered in this book. Governments are involved in regulating and allocating but they are lobbied by, enter partnerships with, and increasingly sub-contract functions to private organizations, including firms. It was shown how susceptible the European Commission was to the blandishments of the roundtable of leading European electronics firms in pressing for a science and technology budget. Governments have innovation advisers from science-based business who may be sufficiently sympathetic to have made party contributions, in return for which they anticipate more than honours for services rendered. They may expect sea-changes in governmental stances on regulatory and fiscal, let alone research-funding budgets. Regional interests will lobby in the way industry does, though less powerfully unless there are prospects of civil disorder.

Governance also involves the micro-management of Knowledge Economies at the level of the cluster. Successful clusters that are Knowledge Economies tend to have private governance of day-to-day collective affairs, and these are for firms much less important in general than their own business management concerns. But the chapter on governance shows just how important venture capitalists can be in cluster management, a feature that has become apparent as they have matured in experience and scale. Thus companies like this may see the advantages of co-operation that this book sees as important to Knowledge Economies, as business assets and evolve private forms of social capital or 'contacts' that are much sought after by start-up and spin-off businesses. They may, effectively, develop their own private clusters like Kleiner Perkins' influential *keiretsu* model in Silicon Valley. In the absence or alongside such market-focused collaboration is that of the regional or local business association or industry council, performing collective services for industry members. They also may help or directly lobby governments for better treatment for their members, like getting the US Food and Drug Administration to open an office in Boston where the biotechnology cluster is on the doorstep, as the Massachusetts Biotechnology Council did.

Governance is thus important to knowledge economies, but so are more informal modes of interaction. Chapter 4 explored some of these, particularly those associated with what are called untraded interdependencies such as trust rather than arm's-length monetary exchange, reputation as a basis of doing business or familiarity from belonging to the same educational, ethnic or professional fraternity, and reliability as a reason for repeat business or even repeat requests for returnable favours. The collective term for this is social capital and some influential authors wonder whether this is the missing ingredient that development theories have been missing since Malthus asked his famous question of Ricardo about the reasons why some economies were rich and others poor. There is widespread evidence that social capital assists educational performance, other things held equal, and Putnam's (1993) work, which turned the perspective towards economic development, presented a cogent argument but not wholly convincing data in support of the thesis that good economic performance is associated with high social capital at regional level. Work that moves beyond Putnam, though, warns of the problem of introversion or too much 'embeddedness' as ultimately a constraint on developmental potential.

Entrepreneurship demands a degree of autonomy from, at least, community-based sources of start-up capital. Ironically, however, as we have seen, actual practices in successful Knowledge Economies are to emphasize the importance of proximity to investment in innovative companies and secure the cluster by investing privatized social capital, represented by the Rolodex, to further integrate it as a private innovation system. A slight doubt can be registered at this point only to arise again in reference to Chapter 6, but as a doubt not a fully-fledged fact. The work of Zucker *et al.* (1998) on biotechnology clusters suggests that industry 'spillovers' like untraded interdependencies do not exist except for trivial exchanges, and that non-trivial ones are much more formal, time-limited, project-focused, hemmed in by confidentiality agreements and often subject to arm's-length exchange contractual agreements. The studies to check this or examine it in other contexts will doubtless occupy many research hours in future years.

The form taken by social capital and its key elements like trust, reciprocity and reputation is a network, and the active process is networking. Economic networks are systems of persons who are nodes linked by communication channels integrated around a shared economic interest. These refer to any aspect of business as well as regular linkages between firms and organizations of relevance to firms, such as public funding agencies, research laboratories, training organizations and the like. In Chapter 5 there is a detailed analysis of networking, including policies that seek to promote it with incentives for firms learning to upgrade in networks. The Danish business network programme is one of the best-known examples. Key to networks is their maintenance through ensuring traffic flows along them, so repeat interactions, follow-ups, feedback and a project focus are valuable impulses in economic networking. Networks function well at distance, because they are often focused on exchange of concrete or codified knowledge like project memoranda, logistical solutions, and ways of taking action to enhance business performance. In most cases they rely on machines and technologies being a digital underpinning, although whether, as Callon (1991) says, PCs or faxes are network members can be disputed. Particular types of network now fundamental to modern governance are policy networks. In national and regional governance these consist of a lead government minister and civil servants in whose field of competence the matter of consequence falls. These are supported by junior ministers from neighbouring departments or civil servants representing them, representatives of economic agencies, trade unions, academia and the private sector. They function by interacting around an agreed agenda, moving from a broadly defined policy aspiration in a series of meetings to produce a set of recommendations for action that have a consistent policy-design. Such recommendations are then implemented by the key minister or an appropriate neighbouring one.

Clusters of all geographically concentrated kinds (there is a weakening argument from Porter, 1990 that they can be aspatial) function through networks of an informal and formal, hard and soft kind involving a variety of interactions from 'untraded interdependencies' to arm's-length exchange and a handshake to a legally enforceable contract. Clusters, though consisting in geographical proximity are generally looser arrangements than networks, with many more potential members, keeping in mind that Silicon Valley has, for example, at least 6,000 firms. Most networks consist of a lot less than a hundred members, and in the Danish programme the average was seven firms per network. Clusters are less project-focused than networks, more open arenas for the kind of interactions mentioned. Of course, project-focused networks may thrive in such proximate settings too. Clusters have risen to prominence over networks because of their high visibility as Knowledge Economies. For decades, policymakers have sought to learn the ingredients of the master cluster of Silicon Valley because of its visibility, growth performance, globally leading innovation position and, nowadays, the abundant evidence that it reproduces those characteristics on a cyclical basis. Other clusters, like Richardson, the two Cambridges, Silicon Alley and Soho, London, capture almost equal attention. All governments want some of the New Economy industries and Knowledge Economy clusters are the most sought-after of all.

In Chapter 6 some key ingredients of successful Knowledge Economies were disclosed. It is held that there is an economy culture denoting New Economy business practices that has some distinctive features, although it is important to note that, in that chapter, it was said that some of those features, like high risk-taking, high competitiveness and a strong growth predisposition could easily fit the era when motoring, flying or music recording had just become possible and the market entry of new producers was very high. Over time consolidation occurred in those clusters, as has happened in the propellant sector of the New Economy 'bubble' dot.com or Internet content businesses. However, workers valuing stock options above wages is a cultural novelty, the idea of true lifelong learning or the 'creative forgetting' of a single hard-won skill in favour of agility in labour markets, and most of all, the enhanced valuation of intelligence as the means by which knowledge acts upon knowledge to create productivity are distinctive. Moreover, the idea that such intelligence is scarce while capital for investment is abundant is not familiar from an Old Economy viewpoint. Clustering is almost explained in that paradox, not least because abundant capital floods to scarce knowledge in a modern, more innovative, but still, ultimately, Schumpeterian swarm intent on disruptive change and creative destruction.

These metaphors can suggest that the New Economy hype has a contaminating effect on prose, but the following case shows that it is indeed a cool description of a certain kind of reality. In the analysis of Northern Ireland's innovation system presented in Chapter 3 (Cooke *et al.*, 2001b) the head of R&D for optical networking in Canadian firm Nortel Networks explained the presence of a major 600-employee research facility on the outskirts of Belfast in simple terms. Optical networking is the technology that organizes digital information flows along fibre optic cable. At various points in each fibre (slimmer than a human hair, as the publicity goes) are mirrors that divert messages to required destinations. Broadband cable, needed to carry video imaging that needs one thousand times the bandwidth of the spoken word, demands complex software engineering to ensure flawless transmission. The technology is also important in genetic screening in biotechnology. The industry, typically, clusters in San Diego, for example, and in the UK in an advanced opto-electronics cluster with the UK's highest cluster location quotient, at St Asaph in Wales.

There is a global shortage of software engineering talent, especially telecommunications software for optical networks. Northern Ireland produces many such engineers, not least because it retains a selective educational system that develops engineering talent the equal of that in any other Nortel location. Of that talent, the industry, dominated by Nortel, Lucent, and Alcatel, has a requirement that 10 per cent should be capable of 'disruptive technological change', that is radical innovation. Though Northern Ireland has some way to go to satisfy the 10 per cent criterion, it produces enough disruptive talent to be attractive as a location for Nortel. The point of the disruptiveness, of course, is to 'blow away the competition', perhaps an unfortunate phrase in the context. This is how representatives of large firms have to think, but another reason for the location decision is that telecommunications software firms are prominent among those starting up in the university incubators, and Nortel will develop supplier arrangements to buy technologies and services locally that it cannot afford to research itself. The Richardson cluster grew around telecommunications technologies to be a major employer of software engineering talent; Figure 3.2 suggests Northern Ireland has many appropriate elements to grow its cluster too, not least regional governance, high-grade research and training, incubation and venture capital.

Which brings us to the question raised in Chapter 7, can clusters, the carriers of Knowledge Economies, be built? Despite the scepticism from many in the worlds of academe and policy, the unequivocal answer of this book is that they can. But, predictably, that statement needs qualification. For a start, built by whom? Clearly, as this book shows over and over again, a combination of a large capital injection to fund basic or applied research, or the presence of one or more of Michael Porter's 'demanding customers' in a place, and venture capital to fund start-ups, can stimulate clustering under circumstances governed by one key condition. The condition is that the demanding customers choose to or can no longer satisfy their own demand by making the innovative products or services they need, and are forced to buy them. Moreover, in Knowledge Economies, the need for knowledge to act upon knowledge means that demand cannot be satisfied by buying 'off the shelf' or codified knowledge, but rather by co-operating with specialist firms to enable tacit knowledge to act upon tacit knowledge to produce innovation, perhaps disruptive innovation.

For Digital Economy firms the demand is from the gargantuan media industry paying commissions for content rather than funding basic research. But in all the circumstances of Knowledge Economy clusters examined in this book, those two distinctive forms of funding input, to research and entrepreneurship, are key.

Now, the real burden of the question is, can public economic development authorities build Knowledge Economy clusters? On this the answer is still positive but less clear-cut. Chapter 7 showed that multi-level governance is complex but can be made to work in support of regional innovation systems and policies. Whether between local and national agencies or supranational to regional agencies, provided there is a governance not merely a government strategy, consensus from private and other non-governmental organizations in the policy network, practical engagement and funding commitment from the governance structure, a process can be put to work. But put to work on what? Key to successful instances of public leadership in governance of Knowledge Economy cluster-building from Austin, Texas, to Oulu, Finland, is the 'triple helix' (after Etkowitz and Leydesdorff, 1997) of industry, government and academia, but unlike their model, rooted in a region or locality with a willingness to build on social capital of the public and the private variety. Thus, once again, we cannot forget the people with the deep pockets and the huge Rolodexes, venture capitalists. Germany has learnt this lesson of cooperative advantage after years of failing with top-down initiatives. Now it has an immature but successfully emerging biotechnology industry, as Finland has a globally dominant mobile telephony industry, closely followed by Sweden, both of which built clusters centred upon university science parks, the 'triple helix', and venture capital. Where private venture capital is not abundant, public venture funds must fill the gap of market failure, something which until 2001, the EU was unaccountably reluctant to allow on spurious grounds of state aids infringing competition rules (on public venture capital, see Doran and Bannock, 2000). On these grounds and until large firms rediscover how to make not buy knowledge and innovation, Knowledge Economies are here to stay and expand for the foreseeable future.

Where next for Knowledge Economies?

There are three visible challenges for Knowledge Economies and those who benefit from them currently and who wish to benefit from them in future. The first of these involves tackling the major regional imbalance that attends Knowledge Economies. The statistics quoted at the outset of this chapter show how much knowledge-based activity has come to dominate modern economies, and the extent to which it concentrates in Knowledge Economy clusters. The ingredients of such clustering are by now clear and consist of five elements, all of which are amenable to policy intervention, although in their detailed functioning non-public actions tend to produce superior results. The first of these interventions is research funding, mostly from the public purse because of well-known risks, indeed, one of the most major of market failures in the modern economy. This obviously applies much more to scientific knowledge and its exploitation than to output from creative industries in the Digital Economy where mostly private contracts will furnish resources. In the Northern Ireland case, despite current changes in organizational structure, the existence of some degree of regional institutional economic autonomy even when the country was ruled directly from London, but conceivably more so with the setting up of a devolved Executive meant policies could be tailored to meet regional economic needs. In particular, the presence of an innovation agency that could make research and innovation funding available to firms and university researchers to work together or separately has been invaluable to the upgrading of some businesses in traditional industries, enabling them to be more globally competitive, as well as assisting start-up firms in university incubators and elsewhere to reach sustainability. As in the USA, which does the same through its Small Business Innovation Research (SBIR) grants, here is an appropriate role for intervention because the investment market is failing regarding this kind of activity. Each regional governance system should have this function to enable it to become a regional innovation system.

Second, investment in new businesses and a systemic method of supporting them with appropriate financial resources and management in the way venture capital companies do are key ingredients. At regional level, albeit in relationship to other levels of governance, there are related requirements. The process by which Scotland proceeded is instructive and, in terms of producing a platform for Knowledge Economy and other clusters, was successful, though it is too early to make an equivalent judgement on the policy output and outcomes side. A political decision to pursue a cluster strategy was taken and a learning process initiated by hiring Monitor, Michael Porter's consultancy, to scope the economy, identify appropriate potential clusters, and train staff at Scottish Enterprise, the economic development agency, to build clusters. These were industries that in some measure already existed, a key factor since it is known from French and Japanese experience with technopoles that it takes a generation before such 'clusters in the desert' show even limited signs of clustering activity (Castells and Hall, 1994; Asheim and Cooke, 1998; Longhi, 1999).

Building a policy network and stakeholder consensus for each planned cluster is a first step, then initiating the process by envisioning the cluster, identifying actors for leadership roles and mobilizing resources for specific expenditure to fill gaps in agreed areas of weakness like supply chains or commercialization facilities quickly follow. An example of the latter is the 'proof of concept' fund enabling scientists to buy out their time to commercialize a discovery.

Third, incubating firms in a manner different from the way many incubators function currently is required. More human capital is necessary than is often available. As we saw in discussing the ways venture capitalists manage their equity investments and, as in the case of a firm like Kleiner Perkins' *keiretsu* or the smaller Belfast investors, encouraging them to inter-trade and engage in clustering practices, start-ups need this kind of attention from the outset. Once they have moved beyond incubation stage, they may require less detailed attention but even firms as globally successful as Sun Microsystems remain members of the KP *keiretsu*. Incubators conceived as real estate are not best suited to encourage networks and cluster-building, not least because firms are likely to be from diverse, possibly incompatible backgrounds. Such set-ups may have their place and some start-up business owners will prefer confidentiality to openness. But customized incubators dedicated to specific sectors like biotechnology, new media or software function successfully as the 55 Broad Street illustration in Chapter 6 showed.

Fourth, venture capital is a key and indispensable ingredient in Knowledge Economy cluster-building. It has been noted that it is conditioned in major ways by geographical proximity as Zook (2000) notes, referring to the Californian penchant for a venturing radius of one-hour's drive because of the need to be able to be 'hands-on' in management of the investment. Where venture capital of the appropriate, incubation-minded kind is subject to market failure, there is a case for public intervention. Doran and Bannock (2000) describe numerous examples of successful public venture capital in US states though at $6 billion the funds are obviously dwarfed by private venture capital, which was over $69 billion for start-ups alone in 2000. California accounted for 40 per cent of this (PWC, 2001). In England, regional development agencies were given venture funding in 2000 by the Treasury (ministry of finance) while in Scotland and Wales such funds already existed. Mention was made earlier of multi-level governance problems at EU level where the policy was blocked on the misunderstanding that they breached competition policy rules. This was, in effect, breached by the long-delayed Northern Ireland case of Viridian, a public–private venture fund with start-up and incubation ambitions, approved in 2001, but in all cases, support is constrained to small, mainly start-up rather than medium-sized businesses.

Finally, because innovation systems of the kind that integrate Knowledge Economies are themselves prone to the same laws of imbalance that this book has been informed by, clustering activity in less favoured regions must also reach further than the regional metropolis. There are numerous ways to achieve this with, once more, policy intervention necessitated due to the market's preference for taking the line of least resistance. One is to ensure that all sub-regions are included in the publicly funded research infrastructure, universities being the most obvious case. This means linkage to specialist institutes based in community colleges for instance, which happens in some Nordic countries and US states like Wisconsin. An innovation of some applicability to rural areas that differs from, say, attaching a biotechnology incubator to an agricultural college, is that found in the new National Botanic Garden in Wales which has incubation facilities for bioscientific businesses as well as plants. The idea was influenced by the tradition in Carmarthenshire of the 'Physicians of Myddfai' a medieval guild of apothecaries. But 'Economic Botany' used to be a rationale for exploration (and colonization), as the presence of such institutes in the mainly urban Botanic Gardens of Australian cities like Adelaide testify. In general, imaginative reconstructions of past practices like these seem beyond markets to recreate so, once again, the role for public sector intervention and funding is clear.

Two other future challenges also involve those who are currently left out of the Knowledge Economy equation. So, the second challenge, for which the foregoing discussion may also be partly relevant, concerns less developed countries. Two illustrations, not dealt with elsewhere in this book, give hope that Knowledge Economies can develop where none existed before. The first comes from Israel, of course not less developed in the classical sense, but with the Palestinian areas economically bereft and a modest ranking in the thirties in the Global Competitiveness Index, not fully developed either. In the 1990s Israel became a globally leading centre for the production of data security software. Firms cluster in Tel Aviv and to a lesser extent Haifa. The knowledge source is military, where encryption expertise that was later exploited for Internet 'firewalls' was first applied. But the key algorithms were developed in the Weizmann Institute of Science. The Camp David peace accords resulted in reductions in Israeli armed forces and military intelligence functions and some of these personnel were responsible for early start-ups in civilian encryption software for electronic mail and the like. An influx of Russian refugees with software and entrepreneurship skills honed in the post-Soviet era coincided with the next phases of growth.

But Israel had no venture capital, only a government research fund rather like SBIR, so Yozma, a public venture capital fund was established. This funded the early start-ups and, crucially, made strong contacts with US venture capital firms in building up its funds and investment syndicates. This was extraordinarily successful in that large numbers of start-up businesses in various aspects of software engineering were able to be funded. Teubal (2000) estimates that Israeli technology start-ups can now be numbered in the thousands. There are over seventy venture capital companies in Israel and over $1 billion available in funding for new ventures. Many of the earlier data security businesses have registered on NASDAQ, the US technology stock market, while some were acquired by US firms. This example of 'Silicon Valley Offshore' was repeated at around the same time in India, particularly Bangalore, and more recently Hyderabad. Bangalore also has a prestigious Institute of Science, built in the days of the Raj. Its university has fourteen engineering colleges and numerous Indian government-owned defence firms like Hindustan Aeronautics were moved there away from trouble-prone border areas. Where aerospace flourishes it is not unknown for the demand for software engineering skills to be high, as occurs in Seattle, and such was the case in Bangalore. However, shortages of such skills in the USA, especially Silicon Valley, led to Indian software engineers becoming 'to-and-fro migrants' as Balasubramanyam and Balasubramanyam (2000) describe them. They learnt about entrepreneurship and venture capital in the USA, some returned to work as software consultants or 'bodyshops' in US firms that had also moved to Bangalore to recruit much-needed labour locally. The bodyshops grew and have become free-standing software firms in their own right. These are success stories, but they are valuable in showing how Knowledge Economies can grow in less developed countries, following exactly the same kind of logic of public research and, eventually, private commercialization of knowledge as occurs in developed countries. The essential features are now well understood so it is for local policy-makers to apply them appropriately in their own settings.

Finally, a generic problem arising from the institutional structures of the Old Economy being locked-in and ill-attuned to the New Economy, but also to the new needs of Old Economy businesses, poses a challenge for policy-makers everywhere. Their mode of assistance to firms was largely designed with inward investment in mind. Not all of this was necessarily foreign investment but include movements from rich to poor regions within countries too. Hence such aids are aids to capital. Grants to offset costs of building new plant and tax relief on the cost of investment in plant and machinery, for example. While these continue to serve a useful function under those circumstances, most governments pursuing a Knowledge Economies strategy also place emphasis on entrepreneurship and endogenous growth. In many countries the previous preponderance of large firms in the employment and GDP indices has been overturned and SMEs provide the larger shares of both. As we have seen in respect of Knowledge Economies, the most-valued service to small technology-intensive firms is management support. It is generally the case that SMEs of all kinds are less than optimally managed, with managers not having a clear idea of where their profit actually comes from, having no time to think about innovation, nor prioritizing exports sufficiently. In the research discussed in Chapter 3 on Northern Ireland, most SMEs (an overwhelming majority of whom are non-innovators) found capital grants of the kind available to them from government inappropriate to their needs because they did not need expensive new machinery but rather management advice. So, for all SMEs, not only the technologically sophisticated ones who, as we have seen, sell a piece of their equity in exchange for Knowledge Economy management advice from venture capital, support for management innovation is much needed. Perhaps an innovative public approach would be to sponsor or implement itself an equivalent service in exchange for equity or loans. In that way the lessons of success learned in Knowledge Economy clusters from New Economy practices and culture could permeate the SME firm population at large, revealing the importance of co-operative advantage for enhancing competitiveness.

Bibliography

Acs, Z. (ed.) (2000) *Regional Innovation, Knowledge and Global Change*, London: Pinter.

Agnew, J. (ed.) (1996) *Political Geography: A Reader*, London: Edward Arnold.

Agnew, J, Livingstone, D, and Rogers, A. (eds) (1996) *Human Geography : An Essential Anthology*, Oxford: Blackwell.

Altenburg, T. and Meyer-Stamer, J. (1999) 'How to promote clusters: policy experiences from Latin America', *World Development*, 27, 1693–714.

Amin, S, (1976) *Unequal Development*, Brighton: Harvester.

Andersen, E. (1992) 'Approaching national systems of innovation', in B. Lundvall (ed.) *National Systems of Innovation*, London: Pinter.

——. (1995) *Evolutionary Economics: Post-Schumpeterian Contributions*, London: Pinter.

Anselin, L., Varga, A. and Acs, Z. (1997) 'Local geographic spillovers between university research and higher technology innovations', *Journal of Urban Economics*, 42, 422–48.

Arrow, K. (1973) *Information and Economic Behaviour*, Stockholm: FSI.

Asheim, B. and Cooke, P. (1998) 'Localized innovation networks in a global economy: a comparative analysis of endogenous and exogenous regional development approaches', in F. Engelstad, G. Brochmann, R. Kalleberg, A. Liera and L. Mjøset (eds), *Comparative Social Research*, 17, 199–240, London: Jai Press Inc.

Argyris, C. and Schon, D. (1978) *Organisational Learning: A Theory of Action Perspective*, Reading, MA: Addison-Wesley.

Arthur, B. (1994) *Increasing Returns and Path Dependence in the Economy*, Ann Arbor, MI: Michigan University Press.

Audretsch, D. and Feldman, M. (1996) 'Knowledge spillovers and the geography of innovation and production', *American Economic Review*, 86, 630–40.

Axelrod, R. (1984) *The Evolution of Cooperation*, London: Penguin.

Balasubramanyam, V. and Balasubramanyam, A. (2000) 'The software cluster in Bangalore', in J. Dunning (ed.) *Regions, Globalization, and the Knowledge-Based Economy*, Oxford: Oxford University Press.

Banfield, E. (1958) *The Moral Basis of a Backward Society*, Glencoe, ILL: Free Press.

Becattini, G. (1989) 'Sectors or districts: some remarks on the conceptual foundations of industrial economics', in E. Goodman and J. Bamford (eds) *Small Firms and Industrial Districts in Italy*, London: Routledge, 123–35.

Belussi, F. (1996) 'Local systems, industrial districts and institutional networks: towards a new evolutionary paradigm of industrial economics', *European Planning Studies*, 4, 1–15.

Bergman, E. (1998) 'Industrial trade clusters in action: seeing regional economies whole', in M. Steiner (ed.) *Clusters and Regional Specialisation*, London: Pion, 92–110.

Bessant, J., Jones, D., Lamming, R. and Pollard, A. (1984) *The West Midlands Automotive Components Industry*, Birmingham: WMCC.

Biggiero, L. (2000) *The Location of Multinationals in Industrial Districts: Knowledge Transfer in Biomedicals* (http//www.uiss.it/faculta/economia/biggiero).

Braczyk, H., Cooke, P. and Heidenreich, M. (eds) (1998) *Regional Innovation Systems*, London: UCL Press.

Braczyk, H., Fuchs, G. and Wolf, H. (eds) (1999) *Multimedia and Regional Economic Restructuring*, London: Routledge.

Bronson, M. (1999) *The Nudist on the Late Shift*, London: Secker and Warburg.

Brusco, S. (1989) 'A policy for industrial districts', in E. Goodman, J. Bamford and P. Saynor (eds) *Small Firms and Industrial Districts in Italy*, London: Routledge.

Brusco, S., Cainelli, G., Forni, F., Franchi, M., Malusardi, A. and Righetti, R. (1996) 'The evolution of industrial districts in Emilia-Romagna', in F. Cossentino, F. Pyke and W. Sengenberger (eds) *Local and Regional Response to Global Pressure: The Case of Italy and its Industrial Districts*, Geneva: International Institute for Labour Studies.

Burt, R. (1992) *Structural Holes: The Social Structure of Competition*, Cambridge, MA: Harvard University Press.

Burton, P. and Smith, R. (1996) 'The United Kingdom', in H. Heinelt and R. Smith (eds) *Policy Networks and European Structural Funds*, Aldershot: Avebury.

Cairncross, F. (1997) *The Death of Distance: How the Communications Revolution Will Change Our Lives*, Boston: Harvard Business School Press.

Callon, M. (1991) 'Techno-economic networks and irreversibility', in J. Law, (ed.) *A Sociology of Monsters: Essays on Power, Technology and Domination*, London: Routledge.

Cappellin, R. (1998) 'The transformation of local production systems: international networking and territorial competitiveness', in M. Steiner (ed.) *Clusters and Regional Specialisation*, London: Pion, 57–80.

Casper, S., Lehrer, M. and Soskice, D. (1999) 'Can high technology industries prosper in Germany? Institutional frameworks and the evolution of the German software and biotechnology industries', *Industry and Innovation*, 6, 5–24.

Castells, M. (1996) *The Rise of the Network Society*, Oxford: Blackwell.

Castells, M. and Hall, P. (1994) *Technopoles of the World: The Making of Twenty-first-century Industrial Complexes*, London: Routledge.

CEC (1995) *The Green Paper on Innovation*, Luxembourg: Commission of the European Communities.

—— (1997a) *The Globalising Learning Economy: Implications for Innovation Policy*, Brussels: Commission of the European Communities.

—— (1997b) *First Action Plan for Innovation*, Luxembourg: Commission of the European Communities.

Chiarvesio, M. (1999) 'Networks without technologies in industrial districts of North East Italy', paper presented at Nordregio Seminar on Cluster Policy, Stockholm, December.

Clancy, P., O'Connell, L. and O'Malley, E. (1998) *Clusters in Ireland*, Dublin: National Economic and Social Council.

Clarke, M (1998) 'Joined-up government', *Public Management and Policy Association Review*, 1.

Coase, R (1937) 'The nature of the firm', *Economica*, 4, 386–405.

Cohen, W. and Levinthal, D. (1989) 'Innovation and learning: the two faces of R&D', *The Economic Journal*, 99, 569–96.

—— (1990) 'Absorptive capacity: a new perspective on learning and innovation', *Administrative Sciences Quarterly*, 35, 128–52.

Coleman, J. (1988) 'Social capital in the creation of human capital', *American Journal of Sociology*, 94, 595–121.

—— (1990) *Foundations of Social Theory*, Cambridge, MA: Harvard University Press.

Commons, J. (1934) *Institutional Economics – Its Place in Political Economy*, New York: Macmillan.

Cooke, P. (1983) 'Labour market discontinuity and spatial development', *Progress in Human Geography*, 7, 543–65.

—— (1993) 'Regional innovation networks: an evaluation of six European cases, *Topos*, 6, 1–30.

—— (1995) 'Review of *Technopolis* by Allen Scott', *International Journal of Urban and Regional Research*, 19, 461–2.

—— (1996) 'Building a twenty-first century regional economy in Emilia-Romagna', *European Planning Studies*, 4, 53–62.

—— (1999) *The German Biotechnology Sector, The Public Policy Impact and Regional Clustering: An Assessment*, Report to DTI, Cardiff: CASS.

—— (2000) 'Learning commercialisation of science: biotechnology and the "new economy" innovation system', *Proceedings of the Danish Research Unit for Industrial Dynamics Summer Conference*, Rebild, Denmark, 15–17 June (in *Industry and Innovation*, 2001).

—— (2001) 'Biotechnology clusters in the UK', *Small Business Economics*, 15, 1–17.

Cooke, P., Bechtle, G., Boekholt, P., de Castro, E., Etxebarria, G., Quevit, M., Schenkel, M., Schienstock, G. and Tödtling, F. (1997) 'Business processes in Regional Innovation Systems in the European Union', paper to EU-TSER Workshop on 'Globalization and the Learning Economy: Implications for Technology Policy', Brussels, April.

Cooke, P., Boekholt, P. and Tödtling, F. (2000) *The Governance of Innovation in Europe*, London: Pinter.

Cooke, P., Clifton, N. and Huggins, R. (2001a) *Competitiveness and the Knowledge Economy: The UK in Global, Regional and Local Context*, Regional Industrial Research Report No. 30, Cardiff: Centre for Advanced Studies, University of Wales.

Cooke, P. and da Rosa Pires, A. (1985) 'Productive decentralisation in three European regions', *Environment and Planning A*, 17, 527–54.

Cooke, P., Etxebarria, G., Morris, J. and Rodrigues, A. (1989) *Flexibility in the periphery: regional restructuring in Wales and the Basque Country*, Regional Industrial Research Report No 3, Cardiff: University of Wales.

Cooke, P and Hughes, G. (1999) 'Creating a multimedia cluster in Cardiff Bay', in H. Braczyk *et al.* (eds) *Multimedia and Regional Economic Restructuring*, London: Routledge.

Cooke, P. and Morgan, K. (1998) *The Associational Economy: Firms, Regions and Innovation*, Oxford: Oxford University Press.

Cooke, P., Morgan, K. and Price, A. (1993) *The Future of the* Mittelstand*: Collaboration Versus Competition*, Regional Industrial Research Report, 13, Cardiff; University of Wales.

Cooke, P., Moulaert, F., Swyngedouw, E., Weinstein, O. and Wells, P. (1992) *Towards Global Localisation*, London: UCL Press.

Cooke, P., Roper, S. and Wylie, P. (2001b) *Regional Innovation Strategy for Northern Ireland*, Belfast: Northern Ireland Economic Council.

Cooke, P., Uranga, M. and Etxebarria, G. (1998) 'Regional systems of innovation: an evolutionary perspective', *Environment & Planning A*, 30 1563–84.

Cooke, P. and Wills, D. (1999) 'Small firms, social capital and the enhancement of business performance through innovation programmes', *Small Business Economics*, 13, 219–34.

Cooke, P., Wilson, R. and Davies, C. (1999a) *Urban Networks in Britain: Concept, Indicators and Analysis*, Working Paper No. 1, ESRC 'Cities: Competitiveness and Cohesion' Research Programme 'Virtual Cities: Urban Networks as Innovative Environments' Project, Cardiff: CASS, Cardiff University.

—— (1999b) *Economic Development Hubs or a Spoke in the Wheel?*, ESRC 'Cities' Programme Working Paper 2, Cardiff: CASS.

Cossentino, F., Pyke, F. and Sengenberger, W. (eds) (1996) *Local and Regional Response to Global Pressure: The Case of Italy and its Industrial Districts*, Geneva: International Institute for Labour Studies.

Craven, D. (2000) 'Aftermarket makers', *Red Herring*, p. 422.

Dahmén, E. (1988) ' "Development Blocks" in industrial economics', *Scandinavian Economic History Review*, 36, 3–14.

Daunton, M. (1998) 'Review of *The Wealth and Poverty of Nations* by David Landes', *Times Higher Education Supplement*, 20 March.

David, P. (1985) 'Clio and the economics of QWERTY', *American Economic Review*, 75, 332–7.

Dawkins, R. (1976) *The Selfish Gene*, Oxford: Oxford University Press.

Deal, T. and Kennedy, A. (1982) *Corporate Cultures*, New York: Addison-Wesley.

de Castro, E. *et al.* (1996) *Regional Innovation Systems Profile of the Portuguese Centro Region*, Aveiro: University of Aveiro, Department of Environment.

de Geus, A. (1988) 'Planning as learning', *Harvard Business Review*, March–April, 70–4.

—— (1997) *The Living Company: Growth: Learning and Longevity in Business*, London: Nicholas Brealey Publishing.

Dei Ottati, G. (1994) 'Cooperation and competition in the industrial district as an organizational model', *European Planning Studies*, 2, 371–92.

—— (1996) 'Economic changes in the district of Prato in the 1980s: towards a more conscious and organized industrial district', *European Planning Studies*, 4, 35–52.

De la Mothe, J. and Paquet, G. (eds) (1998) *Local and Regional Systems of Innovation*, Dordrecht: Kluwer.

Department of Trade and Industry (2001) *Business Clusters in the UK: A First Assessment*, London: DTI.

Department of Trade and Industry and Foreign and Commonwealth Office (1998) *Biotechnology in Germany: Report of an ITS Expert Mission*, London: DTI/FCO.

Dertouzos, M., Lester, R. and Solow, R. (1989) *Made in America*, Cambridge, MA: MIT Press.

Desideri, C. and Santantonio, V. (1996) 'Building a third level in Europe: prospects and difficulties in Italy', *Regional and Federal Studies*, 6, 96–116.

Dohse, D. (1999) *The BioRegio Contest: Results of an Empirical Investigation*, Kiel: Institute of World Economics (mimeo).

—— (2000) 'Technology policy and the regions: the case of the BioRegio contest', *Research Policy*, 29, 1111–33.

Doran, A. and Bannock, G. (2000) 'Publicly sponsored regional venture capital: what can the UK learn from the US experience?', *Venture Capital*, 2, 255–86.

Dosi, G (1988) 'Sources, procedures and microeconomic effects of innovation', *Journal of Economic Literature*, 26, 1120–71.

Drennan, M. (1996) 'The dominance of international finance by London: New York and Tokyo', in P. Daniels and W. Lever (eds) *The Global Economy in Transition*, London: Longman, 352–71.

Dunning, J. (ed.) (2000) *Regions, Globalization, and the Knowledge-Based Economy*, Oxford: Oxford University Press.

Dymski, G. (1996) 'On Krugman's model of economic geography', *Geoforum*, 27, 439–52.

Eastern Region Biotechnology Initiative (1998) *Sourcebook '98*, Cambridge: ERBI.

Edquist, C. (1997) 'Introduction: systems of innovation approaches – their emergence and characteristics', in C. Edquist (ed.) *Systems of Innovation: Technologies, Institutions and Organizations*, London: Pinter.

Edstrom, J. and Eller, M. (1998) *Barbarians Led By Bill Gates*, New York: Holt.

Elliott, D. and Marshall, M. (1989) 'Sector strategy in the West Midlands', in P. Hirst. and J. Zeitlin, J. (eds) *Reversing Industrial Decline?*, Oxford: Berg.

Etkowitz, H. and Leydesdorff, L. (1997) *Universities and the Global Knowledge Economy*, London: Pinter.

Evans, P. (1995) *Embedded Autonomy: States and Industrial Transformation*, Princeton, NJ: Princeton University Press.

Feldman, J. and Klofsten, M. (2000) 'Medium-sized firms and the limits to growth: a case study in the evolution of a spin-off firm', *European Planning Studies*, 8, 631–50.

Feser, E. (1998) 'Old and new theories of industry clusters', in M. Steiner (ed.) *Clusters and Regional Specialisation*, London: Pion, 18–40.

Florence, S. (1958) *Investment, Location and Size of Plant*, Cambridge, Cambridge University Press.

Florida, R. and Kenney, M. (1990) *The Breakthrough Illusion*, New York: Basic Books.

Freeman, C. (1992) *The Economics of Hope*, London: Pinter.

—— (1994) 'Critical survey: the economics of technical change', *Cambridge Journal of Economics*, 18, 463–514.

—— (1995) 'The "National System of Innovation" in historical perspective', *Cambridge Journal of Economics*, 19, 5–24.

Fukuyama, F. (1995) *Trust: The Social Virtues and the Creation of Prosperity*, New York: The Free Press.

—— (1999) *The Great Disruption: Human Nature and the Reconstitution of Social Order*, London: Profile Books.

Galar, R. and Kuklinski, A. (1997) *Regional Profile of Lower Silesia*, report to EU-TSER Project 'Regional Innovation Systems: Designing for the Future', University of Warsaw, Euroreg.

Ganesan, G. (1994) 'Determinants of long-term orientation in buyer-seller relationships', *Journal of Marketing*, 58, 225–49.

Gapper, J. and Denton, N. (1997) *All that Glitters: The Fall of Barings*, London: Penguin.

Gelsing, L. and Knop, P. (1991) *Status of the Network Programme: The Results from a Questionnaire Survey*, report prepared for the National Agency for Industry and Trade, Copenhagen.

Gereffi, G. (1996) 'Commodity chains and regional divisions of labour in East Asia', *Journal of Asian Business*, 12, 75–112.

—— (1999) 'International trade and industrial upgrading in the apparel commodity chain', *Journal of International Economics*, 48, 37–70.

Giesecke, S. (1999) *Determinants of Successful S&T Policy in a National System of Innovation*, Vienna: Economics University (mimeo).

Gilbert, G. (1994) 'The finances of the French regions in retrospect', *Regional Politics and Policy*, 4, 33–50.

Goodman, E., Bamford, J. and Saynor, P.(1989) *Small Firms and Industrial Districts in Italy*, London: Routledge.

Gordon, D. (1988) 'The global economy: new edifice or crumbling foundations?', *New Left Review*, 168, 24–64.

Gordon, I. and Lawton Smith, H. (1998) *Economic Structures and Clusters in the Thames Valley*, Slough: Thames Valley Partnership.

Grabher, G. (1993) 'The weakness of strong ties: the lock-in of regional development in the Ruhr area', in G. Grabher (ed.) *The Embedded Firm: on the Socioeconomics of Industrial Networks*, London: Routledge.

Grande, E. (1996) 'The state and interest groups in a framework of multi-level decision-making: the case of the European Union', *Journal of European Public Policy*, 3, 318–38.

Granovetter, M. (1973) 'The strength of weak ties', *American Journal of Sociology*, 78, 1360–80.

—— (1985) 'Economic action and social structure: the problem of embeddedness', *American Journal of Sociology*, 91, 481–510.

Gray, A. (ed.) (1997) *International Perspectives on the Irish Economy*, Dublin: Indecon.

Grote, J. (1993) 'Diseconomies in space: traditional sectoral policies of the EC, the European Technology Community and their effects on regional disparities', in R. Leonardi (ed.) *The Regions and the European Community*, London: Frank Cass.

Hall, P. and Markussen, A. (eds) (1985) *Silicon Landscapes*, London: Allen and Unwin.

Hakansson, H. (ed.) (1987) *Industrial Technological Development: A Network Approach*. London: Croom Helm.

—— (1989) *Corporate Technological Behaviour: Co-operation and Networks*, London: Routledge.

Halaris, G., Kyrtsoudi, M., Nikolaidis, E. and Philippopolou, P. (1991) *Archipelago Europe*, FAST Paper 304, Brussels: CEC.

Harrison, B. (1994) *Lean and Mean: The Changing Landscape of Corporate Power in the Age of Flexibility*, New York: Basic Books.

Harvie, C. (1994) *The Rise of Regional Europe*, London: Routledge.

Hatch, R. (1988) *Flexible Manufacturing Networks: Cooperation for Competitiveness in a Global Economy*, Washington, DC: Corporation for Enterprise Development.

Heidenreich, M. (1996) 'Beyond flexible specialization: the rearrangement of regional production orders in Emilia-Romagna and Baden-Württemberg', *European Planning Studies*, 4, 401–20.

Hendry, C. and Brown, J. (1998) *The Development and Performance of Opto-Electronics in the UK, Germany and the USA*, Report to the Welsh Development Agency, London: City University Business School.

Henry, N. and Pinch, S. (1997) *A Regional Formula for Success? The Innovative Region of Motor Sport Valley*, Birmingham: Dept. of Geography, Birmingham University.

—— (1999) 'Placing a knowledge community: a case study of "motor sport valley"', in J. Bryson, N. Henry and J. Pollard (eds) *Knowledge, Space, Economy*, London: Routledge.

Henton, D., Melville, J. and Walesh, K. (1997) *Grass Roots Leaders for a New Economy*, San Francisco: Jossey-Bass.

Heydebrand, W. (1999) 'Multimedia networks, globalization and strategies of innovation: the case of Silicon Alley', in H. Braczyk, G. Fuchs and H. Wolf (eds) *Multimedia and Regional Economic Restructuring*, London: Routledge.

Hirschman, A. (1958) *The Strategy of Economic Development*, New Haven, CT: Yale University Press.

Hirst, P. (1994) *Associative Democracy*, Amherst, MA: University of Massachusetts Press.

Hodgson, G. (1995) *Evolutionary and Competence-Based Theories of the Firm*, Cambridge: The Judge Institute of Management Studies, University of Cambridge.

—— (1996) 'Varieties of capitalism and varieties of economic theory', *Review of International Political Economy*, 3, 381–434.

Hooghe, L. (ed.) (1996) *Cohesion Policy and European Integration*, Oxford: Clarendon Press.

Huggins, R. (1998) 'Local business co-operation and Training and Enterprise Councils: the development of inter-firm networks', *Regional Studies*, 32, 813–26.

—— (2000) *The Business of Networks: Inter-firm Interaction, Institutional Policy and the TEC Experiment*, Aldershot: Ashgate.

Hughes, G. (1998) *The Danish Business Networking Programme: An Assessment* (mimeo).

Humphrey, J. and Schmitz, H. (1998) 'Trust and inter-firm relations in developing and transition economies', *The Journal of Development Studies*, 34, 32–61.

—— (2000) *Governance and Upgrading: Linking Industrial Cluster and Global Value Chain Research*, Working Paper No. 120, Institute of Development Studies, Sussex University.

Imrie, R.F. (1989) 'Industrial change in a local economy: the case of Stoke-on-Trent', unpublished PhD, University of Wales, Cardiff.

Isard, W. (1956) *Location and Space Economy*, New York: John Wiley.

Jackson, T. (1998) *Inside Intel*, London: HarperCollins.

Jacobs, J. (1961) *The Death and Life of Great American Cities*, Harmondsworth: Penguin.

—— (1969) *The Economy of Cities*, New York: Random House.

—— (2000) *The Nature of Economies*, New York: Vintage.

Jaffe, A., Trajtenberg, M. and Henderson, R. (1993) 'Geographic localization of knowledge spillovers as evidenced by patent citations', *Quarterly Journal of Economics*, 108, 577–98.

Jeffery, C. (ed.) (1996) 'Farewell the third level? The German *Länder* and the European Policy Process', *Regional and Federal Studies*,6, 56–75.

Johannisson, B. (1987) 'Entrepreneurship in a corporatist state: the case of Sweden', in R. Goffee and R. Scase (eds) *Entrepreneurship in Europe: The Social Processes*, London: Croom Helm.

Johanson, J. (1991) *Interfirm Networks in Swedish Industry*, Department of Business Economics, University of Uppsala, Sweden.

Johanson, J. and Mattsson, L. (1985) 'Marketing investments and market investments in industrial networks', *International Journal of Research in Marketing*, 2, 185–95.

Johnson, B. (1992) 'Institutional learning', in B. Lundvall (ed.) *National Systems of Innovation: Towards a Theory of Innovation and Interactive Learning*, London: Pinter.

Jüssila, H. and Segerståhl, B. (1997) 'Technology Centres as Business Environments in Small Cities', *European Planning Studies*, 5, 371–84.

Kaplan, D. (1999)*The Silicon Boys and Their Valley of Dreams*, New York: Morrow.

Keck, O. (1993) 'The national system for technical innovation in Germany', in R. Nelson (ed.) *National Innovation Systems: A Comparative Study*, Oxford: Oxford University Press.

Keeble, D. and Nachum, L. (2001) *Why Do Business Service Firms Cluster?*, Working Paper No. 194, Cambridge: ESRC Centre for Business Research, Cambridge University.

Keeble, D. and Wilkinson, F. (1999) 'Collective learning processes, networking and institutional thickness in the Cambridge region', *Regional Studies*, 33, 319–32.

Kehoe, L. (2001) 'E-business to the rescue of the Valley', *Financial Times*, 21 February, 14.

Kellner, D. (1988) 'Postmodernist aestheticism: a new moral philosophy?', *Theory Culture and Society*, 5, 239–70.

Kelly, K. (1998) *New Rules for the New Economy*, London: Fourth Estate.

Kerremans, B. and Beyers, J. (1996) 'The Belgian sub-national entities in the European Union: second or third level players?', *Regional and Federal Studies*, 6, 41–55.

König, G. (1998) 'Nurturing biotech in the regions', *Pharmaceutical Forum*, 9–11.

Krugman, P. (1991) *Geography and Trade*, Cambridge, MA and London: MIT Press.

—— (1994) 'The myth of Asia's miracle', *Foreign Affairs*, Nov–Dec, 63–75.

—— (1995) *Development, Geography and Economic Theory*, Cambridge and London: MIT Press.

Kuhn, T. (1962) *The Structure of Scientific Revolutions*, Chicago: University of Chicago Press.

Landes, D (1998) *The Wealth and Poverty of Nations*, London: Abacus.

Leadbeater, C. (2000) *Living on Thin Air: The New Economy*, London: Viking.

Leadbeater, C. and Oakley, K. (1999) *The Rise of the Knowledge Entrepreneur*, London: Demos.

Lampulo, E. (1996) *The Oulu Medipolis*, Oulu: Medipolis.

Langlois, R. (ed.) (1993) *Economics as a Process*, Cambridge: Cambridge University Press.

Langlois, R. and Robertson, P. (1995) *Firms, Markets and Economic Change*, London: Routledge.

Lazaric, N. and Lorenz, E. (eds) (1998) *Trust and Economic Learning*, Cheltenham: Edward Elgar.

Lewis, J., Disney, S., Gillingham, R. and Childerhouse, P. (1998) *Quick Scan Report, Lucas Ignition & Components*, EPSRC Innovative Manufacturing Initiative Final Report on Supply Chain 2001 Study, Swindon: Engineering and Physical Sciences Research Council.

Lewis, M. (2000) *The New New Thing: A Silicon Valley Story*, New York: Norton.

Lipietz, A. (1987) *Mirages and Miracles: The Crises of Global Fordism*, London: Verso.

Lindholm Dahlstrand, Å. (1997) 'Entrepreneurial spin-off enterprise in Göteborg, Sweden', *European Planning Studies*, 5, 659–74.

Longhi, C. (1999) 'Networks, collective learning and technology development in innovative high technology regions: the case of Sophia-Antipolis', *Regional Studies*, 33, 333–42.

Longhi, C. and Quéré, M. (1993) 'Innovative networks and the technopolis phenomenon: the case of Sophia-Antipolis', *Environment and Planning C: Government and Policy*, 11, 317–30.

Lösch, A. (1952) *The Economics of Location*, New Haven, CT: Yale University Press.

Lowenstein, R. (2000) *When Genius Failed: The Rise and Fall of Long-term Capital Management*, New York: Random House.

Lucas, G. (1988) 'On the mechanics of economic development', *Journal of Monetary Economics*, 22, 3–42.

Luhmann, N. (1979) *Trust and Power*, Chichester: Wiley.

Lundvall, B. (ed.) (1992) *National Systems of Innovation: Towards a Theory of Innovation and Interactive Learning*, London: Pinter.

Lundvall, B. and Johnson, B. (1994) 'The learning economy', *Journal of Industry Studies*, 1, 23–41.

Machin, D. and Smyth, R. (1969) *The Changing Structure of the British Pottery Industry*, University of Keele, Dept. of Economics.

Maillat, D. (1995) 'Territorial dynamic innovative milieux and regional policy', *Entrepreneurship and Regional Development*, 7, 157–65.

Makó, C., Novoszath, A. and Kuczi, T. (1997) *Organisational Innovation in Féjer region, Hungary*, report to EU-TSER project, 'Regional Innovation Systems: Designing for the Future', Hungarian Academy of Sciences, Budapest.

Malecki, E. (1991) *Technology and Economic Development*, London: Longman.

—— (2000) *The Internet: Its Economic Geography and Policy Implications* (mimeo).

Malecki, E. and Tootle, D. (1996) 'The role of networks in small firm competitveness', *International Journal of Technology Management*, 11, 43–57.

Malmberg, A. and Maskell, P. (1997) 'Towards an explanation of regional specialization and industry agglomeration', *European Planning Studies*, 5, 25–42.

Mariussen, Å., Asheim, B., Chiarvesio, M., Cooke, P., Grabher, G., Nauwelaers, C., Tödtling, F. and Wintjes, R. (2000) *Exploitation of Knowledge Governance* (mimeo).

Marks, G., Scharpf, F., Schmitter, P. and Streeck, W. (1996) *Governance in the European Union*, London: Sage.

Marquardt, M. and Sashkin, M. (1995) *Building the Learning Organization*, New York: McGraw-Hill.

Marshall, A. (1916) *Principles of Economics*, London: Macmillan

—— (1919) *Industry and Trade*, London: Macmillan.

—— (1962) *Elements of Economics of Industry*, New York: Macmillan.

Martin, R. (1999) 'The new "geographical turn" in economics: some critical reflections', *Cambridge Journal of Economics*, 23, 65–91.

Martin, R. and Sunley, P. (1996) 'Paul Krugman's geographical economics and its implications for regional development theory: a critical assessment', *Economic Geography*, 72, 259–92.

Maskell, P., Eskelinen, H., Hannibalsson, I., Malmberg, A. and Vatne, E. (1998) *Competitiveness, Localised Learning and Regional Development: Specialisation and Prosperity in Small Open Economies*, London: Routledge.

Mazey, S., (1995) 'French regions and the European Union', in J. Loughlin and S. Mazey (eds) *The End of the French Unitary State: Ten Years of Regionalization in France*, London: Frank Cass.

Micklethwait, J. and Wooldridge, A. (2000) *For a Perfect Economy*, London: William Heinemann.

Moody, F. (1995) *I Sing the Body Electronic: A Year with Microsoft on the Multimedia Frontier*, London: Coronet.

Morass, M, (1996) 'Austria: the case of a federal newcomer in European Union politics', *Regional and Federal Studies*, 6, 76–95.

Myrdal, G. (1957) *Economic Theory and Underdeveloped Regions*, London: Duckworth.

Nachum, L. and Keeble, D. (2000) *Foreign and Indigenous Firms in the Media Cluster of Central London*, Working Paper No. 154, Cambridge: Cambridge University, Centre for Business Research.

Nelson, R. (ed.) (1993) *National Innovation Systems: A Comparative Analysis*, Oxford: Oxford University Press.

Nelson, R. and Winter, S. (1982) *An Evolutionary Theory of Economic Change*, Cambridge, MA: Harvard University Press.

NESC (1998) *Sustaining Competitive Advantage*, Dublin: The National Economic and Social Council.

NIEC (2000) *The Capabilities and Innovation Perspective: The Way Ahead in Northern Ireland* (author M. Best), Research Monograph 8, Belfast: Northern Ireland Economic Development Council.

Nonaka, I. and Reinmöller, P. (1998) 'The legacy of learning: toward endogenous knowledge creation for Asian economic development', *Wissenschaftszentrum Berlin Jahrbuch*, Berlin: WZB.

North, D. (1993) 'Institutions and economic performance', in U. Mäki, B. Gustafsson and C. Knudsen (eds) *Rationality, Institutions and Economic Methodology*, London: Routledge.

Norton, R. (2000) *Creating the New Economy: The Entrepreneur and the US Resurgence*, Cheltenham, Edward Elgar.

O'Donnell, R. (1998) 'Post-Porter: exploring policy for the Irish context "Sustaining Competitive Advantage"', in NESC (ed.) *Sustaining Competitive Advantage*, Dublin: National Economic and Social Council.

OECD (1999) *Science, Technology and Industry Scoreboard 1999: Benchmarking Knowledge-based Economies*, Paris: OECD.

Ohlin, B. (1933) *Interregional and International Trade*, Cambridge, MA: Harvard University Press.

Orsenigo, L. (2000) 'The (failed) evolution of a biotechnology district: the case of Milan', paper to international workshop on 'Comparing the Development of Biotechnology Clusters', Stuttgart, January.

Osborne, D. (1988) *Laboratories of Democracy*, Boston: Harvard Business School Press.

Osborne, D. and Gaebler, T. (1992) *Reinventing Government: How the Entrepreneurial Spirit is Transforming the Public Sector*, Reading, MA: Addison-Wesley.

Ostrom, E. (1992) 'Community and the endogenous solution of commons problems', *Journal of Theoretical Politics*, 4, 343–51.

Padley, H and Pugh, G. (2000) 'The pottery industry: exploding the myths and charting the way', in I. Jackson, S. Leech, M. O'Keefe and L. Trustrum (eds) *Ceramic Ambitions and Strategic Directions*, York: York Publishing Services.

Pavlik, J. (1999) 'Content and economics in the multimedia industry: the case of New York's Silicon Alley', in H. Braczyk, G. Fuchs and H. Wolf (eds) *Multimedia and Regional Economic Restructuring*, London: Routledge.

Perroux, F. (1950) 'Economic spaces', reprinted in F. Perroux (1969) *The Twentieth Century Economy*, Paris: Presses Universitaires (in French), 25–39.

Pilling, D. (2000) 'Where science and Mammon collide', *Financial Times*, 21 March, 17.

Piore, M. and Sabel, C. (1984) *The Second Industrial Divide*, New York: Basic Books.

Polanyi, K. (1944) *The Great Transformation*, Boston: Beacon Press.

Polanyi, M (1966) *The Tacit Dimension*, London: Routledge and Kegan Paul.

Popper, K. (1965) *The Open Society and its Enemies*, London: Routledge.

Porter, M. (1990) *The Competitive Advantage of Nations*, New York: The Free Press.

—— (1998) *On Competition*, Boston: Harvard Business School Press.

Porter, M., Takeuchi, H. and Sakakibara, M. (2000) *Can Japan Compete?*, London: Macmillan.

Putnam, R. (1993) *Making Democracy Work: Civic Traditions in Modern Italy*, Princeton, NJ: Princeton University Press.

PWC (2001) *Money Tree Survey*, London: Pricewaterhouse Coopers Ltd.

Pyke, F. (1998) *Promoting Enterprise through Networked Regional Development*, Vienna: UNIDO.

—— (2000) *The Silk Industry in Macclesfield* (mimeo).

Pyke, F., Becattini, G. and Sengenberger, W. (eds) (1990) *Industrial Districts and Inter-firm Co-operation in Italy*, Geneva: International Institute for Labour Studies.

Pyke, F. and Sengenberger, W. (eds) (1992) *Industrial Districts and Local Economic Generation*, Geneva: Industial Institute for Labour Studies.

Rehfeld, D. (1995) 'Disintegration and reintegration of production clusters in the Ruhr area', in P. Cooke, (ed.) *The Rise of the Rustbelt*, London: UCL Press.

Ricardo, D. ([1817] 1967) *Principles of Political Economy and Taxation*, Harmondsworth: Penguin.

Richardson Chamber of Commerce (2000) *The Telecom Corridor in Richardson, Texas*, Richardson, TX: Chamber of Commerce.

Ridley, M. (1997) *The Origins of Virtue*, London: Penguin.

Robertson, P. and Langlois, R. (1995) 'Innovation, networks and vertical integration', *Research Policy*, 24, 543–62 .

Romer, P. (1990) 'Endogenous technical change', *Journal of Economic Literature*, 98, 338–54.

—— (1994) 'The origins of endogenous growth', *Journal of Economic Perspectives*, 8, 3–22.

Rosenfeld, S. (1990) *Technology, Innovation and Regional Development: Lessons from Italy and Denmark*, Washington, DC: The Aspen Institute.

—— (1995) *Industrial Strength Strategies*, Washington, DC: Aspen Institute.

—— (1996) 'Does cooperation enhance competitiveness? Assessing the impacts of inter-firm collaboration', *Research Policy*, 25, 247– 63.

—— (1997) 'Bringing business clusters into the mainstream of economic development', *European Planning Studies*, 5, 3–24.

—— (2000) *learning.now: Skills for the Information Economy*, Washington, DC: Community College Press.

Ruigrok, W. and Van Tulder, R. (1995) *The Logic of International Restructuring*, London: Routledge.

Russ, A. (1998) 'Multimedia: a new industry in a global city', unpublished MSc, Department of Geography, King's College, London.

Russo, M. (1985) 'Technical change and the industrial district: the role of interfirm relations in the growth and transformation of ceramic tile production in Italy', *Research Policy*, 14: 329–43.

Sabel, C. (1992) 'Studied trust: building new forms of co-operation in a volatile economy', in F. Pyke and W. Sengenberger (eds) *Industrial Districts and Local Economic Regeneration*, Geneva: IILS, 215–50.

Sako, M (1992) *Prices, Quality and Trust*, Cambridge: Cambridge University Press.

Saxenian, A. (1994) *Regional Advantage: Culture and Competition in Silicon Valley and Route 128*, Cambridge, MA: Harvard University Press.

Schamp, E.W. (2000) *Vernetzte Produktion. Industriegeographie aus Institutioneller Perspektive*, Darmstadt: Wissenschaftiliche Buchgesellschaft.

Schmitz, H. (1998) *Responding to Global Competitive Pressure: Local Cooperation and Upgrading in the Sinos Valley, Brazil*, Working Paper No. 82, Institute of Development Studies, University of Sussex.

Schuller, T. and Field, J. (1998) 'Social capital, human capital, and the learning society,' *Journal of Life Long Education*, 17, 226–35.

Schumpeter, J. ([1928] 1951) *Essays on Economic Topics*, ed. R. Clemence, New York: Kennikat.

—— (1934) *The Theory of Economic Development*, Cambridge, MA: Harvard University Press.

—— ([1942] 1975) *Capitalism, Socialism and Democracy*, New York: Harper Torchbooks.

—— (1951) *The Theory of Economic Development: An Inquiry into Profits, Capital, Credit, Interest and the Business Cycle*, Cambridge, MA: Harvard University Press.

Senge, P. (1995) *The Fifth Discipline*, New York: Doubleday.

Shapira, P. (1998) *The Evaluation of USNet: Overview of Methods, Results and Implications*, Atlanta, GA: School of Public Policy – Institute of Technology.

Smith, A. ([1776] 1976) *An Inquiry into the Nature and Causes of the Wealth of Nations*, Oxford: Clarendon Press.

Smith, K. (2000) 'What is the "knowledge economy"? Knowledge-intensive industries and distributed knowledge bases', paper prepared for European Commission project on 'Innovation Policy in a Knowledge-Based Economy', STEP Group, Oslo, Norway.

Solow, R. (1957) 'Technical change and the aggregate production function', *Review of Economics and Statistics*, 39, 312–20.

Steiner, M. (ed.) (1998) *Clusters and Regional Specialisation*, London: Pion.

Sternberg, R. (2001) 'Internet domains and the innovativeness of cities/regions – evidence from Germany and Munich', paper to ESRC Workshop 'Innovation and Competitive Cities in the Global Economy', Worcester College, Oxford University, 28–30 March.

Stevens, J. (1989) 'Integrating the supply chain', *International Journal of Physical Distribution and Materials Management*, 623–45.

Stigler, G. (1951) 'The division of labour is limited by the extent of the market', *Journal of Political Economy*, 69, 213–25.

Storper, M. (1995) 'The resurgence of regional economies, ten years after: the region as a nexus of untraded interdependencies', *European Urban and Regional Studies*, 2, 191–221.

Sturm, R. (1998) 'Multi-level politics of regional development in Germany', *European Planning Studies*, 6, 525–36.

Swann, P., Prevezer, M. and Stout, D. (eds) (1998) *The Dynamics of Industrial Clustering*, Oxford: Oxford University Press.

Tang, P. (1999) 'The South-East England high tech corridor: not quite Silicon Valley yet', in H. Braczyk, G. Fuchs and H. Wolf (eds) *Multimedia and Regional Economic Restructuring*, London: Routledge.

Teubal, M. (2000) *The Systems Perspective to Innovation and Technology Policy: Theory and Selected Topics*, Jerusalem: Hebrew University, Institute of Israel Studies (mimeo).

Thomas, K. (2000) 'Creating regional cultures of innovation? The regional innovation strategies in England and Scotland', *Regional Studies*, 34, 190–8.

Tödtling, F. and Sedlacek, S. (1997) 'Regional economic transformation and the innovation system of Styria', *European Planning Studies*, 5, 43–64.

Towill, D. (1997) 'The seamless supply chain – the creditor's strategic advantage', *International Journal of Technology and Management*, 13, 37–56.

Trigilia, C. (1992) 'Italian industrial districts: neither myth nor interlude', in F. Pyke and W. Sengenberger (eds) *Industrial District and Local Economic Regeneration*, Geneva: International Institute of Labour Studies.

Varaldo, R. and Ferrucci, L. (1996) 'The evolutionary nature of the firm within industrial districts', *European Planning Studies*, 4, 27–34.

Veblen, T. (1899) *The Theory of the Leisure Class: An Economic Study of Institutions*, New York: Macmillan.

Vernon, R. (1966) 'International investment and international trade in the product cycle', *Quarterly Journal of Economics*, 80, 190–207.

Weber, A. (1928) *The Theory of the Location of Industries*, Chicago: University of Chicago Press.

Weber, M. (1930) *The Protestant Ethic and the Spirit of Capitalism*, London: Allen & Unwin.

White, C. (1997) 'Catalonia's pocket-sized multinationals', *Financial Times*, 7 January.

Williams, G. (2000) 'The digital value chain and economic transformation: rethinking regional development in the "new economy"', *Contemporary Wales*, 13, 94–115.

Williams, R. (1983) *Towards 2000*, London: Chatto and Windus.

Willis, T. (1999) *Oxfordshire Motor Sport Forum*, A report on the actions of the Oxfordshire Motorsport Forum between February 1996 and March 1999 for Heart of England Training and Enterprise Council, Oxford: Oxford Innovation.

Witt, U. (1991) 'Reflections on the present state of evolutionary economic theory', in G. Hodgson and E. Screpanti (eds) *Rethinking Economics: Markets, Technology and Economic Evolution*, Cheltenham: Edward Elgar.

Womack, J., Jones, D. and Roos, J. (1990) *The Machine That Changed the World*, London: Macmillan.

Woolcock, M. (1998) 'Social capital and economic development: towards a theoretical synthesis and policy framework', *Theory and Society*, 27, 151–208.

Wyles, J., Kimbel, J. and Wilson, A. (1993) 'Birds, behaviour and anatomical evolution', *Proceedings of the National Academy of Sciences*, July.

Zook, M. (2000) 'Grounded capital: venture capital's role in the clustering of Internet firms in the US', paper prepared for Association of Collegiate Schools of Planning 2000 Conference, Atlanta, GA, November.

Zucker, L., Darby, M. and Armstrong, J. (1998) 'Geographically localised knowledge: spillovers or markets?', *Economic Inquiry*, 36, 65–86.

Name index

Subject index